Quality of Higher Education: Organizational and Educational Perspectives

Innovation and Change in Education – A Cross-cultural Perspective

Volume 3

Series Editor

Xiangyun Du
Aalborg University, Denmark

Nowadays, educational institutions are being challenged when professional competences and expertise become progressively more complex. This is mainly because problems are more technology-bounded, unstable and ill-defined with the involvement of various integrated issues. To solve these problems, it requires interdisciplinary knowledge, collaboration skills, innovative thinking among other competences. In order to facilitate students with the competences expected in professions, educational institutions worldwide are implementing innovations and changes in many aspects.

This book series includes a list of research projects that document innovation and change in education. The topics range from organizational change, curriculum design and innovation, pedagogy development, to the role of teaching staff in the educational change process, and quality issues, among others. A cross-cultural perspective is studied in this book series that includes two layers. First, research contexts in these books include different countries with various educational traditions, systems and societal backgrounds. Second, the impact of professional and institutional cultures such as engineering, medicine and health, and teachers' education are also taken into consideration in these research projects.

For a list of other books in this series, please visit http://riverpublishers.com

Quality of Higher Education: Organizational and Educational Perspectives

A Study in the Chinese Context

Yihuan Zou
*Department of Learning and Philogophy,
University of Aalborg, Denmark*

Aalborg

ISBN 978-87-92982-55-1 (hardback)
ISBN 978-87-92982-56-8 (ebook)

Published, sold and distributed by:
River Publishers
P.O. Box 1657
Algade 42
9000 Aalborg
Denmark

Tel.: +45369953197
www.riverpublishers.com

All rights reserved © 2013 River Publishers

No part of this work may be reproduced, stored in a retrieval system, or transmitted in any form or by any means, electronic, mechanical, photocopying, microfilming, recording or otherwise, without prior written permission from the Publisher.

Contents

Acknowledgements vii

Preface ix

Clarification of Terms xi

1 Introduction 1

2 Quality of higher education in the Chinese context 5
  2.1 Education in the Chinese context 5
  2.2 A brief introduction to higher education in China.....................7
  2.3 Expansion and quality assurance in Chinese higher
       education 12

3 Perspectives to study quality of higher education 19
  3.1 Quality assurance in higher education....................................19
  3.2 Quality of higher education – the concept..............................21
  3.3 Perspectives to approach quality of higher education 22
  3.4 Mapping current research with the perspectives from
       the policy, organization and education...................................27
  3.5 Researching actors' perceptions and concerns about
       quality of higher education...31
  3.6 The focus of this project...33

4 Researching quality of higher education in China 35
  4.1 Methodological assumptions and choices 35
  4.2 Overall design of the study..36
  4.3 Content analysis of institutional self-evaluation reports 38
  4.4 Teaching staff interview and student focus group..................40
  4.5 Methodological reflections...48

5 The organizational perspective   51
    5.1 The contents of the self-evaluation reports 51
    5.2 Document-listing and awards as visible forms of quality 56
    5.3 Higher education organizations and their environments 58
    5.4 Summary of quality from the self-evaluation report
        point of view 60

6 The teaching staff perspective   63
    6.1 Teachers and Teaching 63
    6.2 Students and learning 89
    6.3 Institutions and higher education 96
    6.4 Summary of quality from the teaching staff point of
        view 100

7 The student perspective   103
    7.1 Students and learning 103
    7.2 Teachers and teaching 132
    7.3 Institutions and higher education 137
    7.4 Summary of quality from the student point of view 143

8 Exploring the tensions in quality of higher education   145
    8.1 Summary of perceptions of quality of higher education 145
    8.2 The orientations and tensions in the pursuit of quality 146
    8.3 The dilemma between efficiency and sustainable
        development 154

9 Conclusion   157
    9.1 Summary of this study 157
    9.2 Answers to research questions and implications 158
    9.3 The complexity of quality in higher education 160
    9.4 Toward a more inclusive understanding of quality in
        higher education 165

A Profile of teaching staff participants 167

B Profile of student focus group participants 169

Bibliography   171

About the author   181

Acknowledgements

Grateful acknowledgement is made to Taylor & Francis Group (http://www.tandfonline.com) for the permission to reproduce here my findings from institutional self-evaluation reports, which were previously published as following:

Zou, Y., Du, X., & Rusmussen, P. (2012). Quality of higher education: Organisational or educational? A content analysis of Chinese university self-evaluation reports. *Quality in Higher Education*, 18(2), 169-184.

I would also like to express my special thanks to those organizations and people who offered me precious help with the research project on which this book is based. Among them are:

China Scholarship Council, which provided most of the funds for the research project;

Aalborg University, which hosted the research project;

All the participants who shared with me their perceptions and experience, which are essential for the discussions in this book; and finally, Prof. Xiangyun Du and Prof. Palle Rasmussen, who offered me invaluable comments on the research project.

Preface

This book attempts to contribute to a more inclusive understanding of the increasingly concerned issue of quality in higher education. To achieve this, it has firstly constructed a three-perspective framework to approach the quality issue in higher education, i.e. the policy perspective, the organizational perspective and the educational perspective, of which the policy perspective is currently prominent in research literature. With the aim of supplementing the policy perspective on the quality of higher education, it endeavors to address the quality issue from organizational and educational perspectives. The assumption is that, with multiple perspectives that bring macro policy concerns of quality in parallel with the meso organizational behavior and micro teaching and learning activities, a more informed picture about quality in higher education could be formed.

Clarification of Terms

Higher education, university education

Higher education in this book refers to undergraduate education that leads to a bachelor's degree. More specifically, in the Chinese context it refers to the four-year undergraduate education (or five-year for some medical programs) provided at regular higher education institutions (Chapter 2 will offer more information about the higher education system in China). Sometimes, it is also referred to as university education, especially in spoken Chinese.

Higher education institution, university

In this book, I do not distinguish between a higher education institution and a university: both of them refer to those post-secondary institutions that offer undergraduate education with a bachelor's degree. In the Chinese context, they refer to regular higher education institutions.

Quality assurance, evaluation

There are several terms related to the quality practice in higher education. In this book, 'quality assurance' refers to all kinds of work related to specialized quality management of higher education, which includes different approaches such as accreditation, audit, evaluation, benchmarking, etc. And 'evaluation' refers to the review and making judgments of the work of higher education institutions. Specifically, in this book the discussion on institutional evaluation in China is about the undergraduate teaching evaluation of regular higher education institutions initiated by the Chinese Ministry of Education and conducted between 2003 and 2008 (Chapter 2 will provide more information about this).

Self-evaluation report

Self-evaluation report or institutional self-evaluation report in this book refers specifically to the self-evaluation report produced by the institutions during the 2003–2008 undergraduate teaching evaluation in China. These reports form the basis of the analysis of the institutional point of view.

Evaluation Plan

Evaluation Plan in this book refers to the '*Plan for Undergraduate Teaching Evaluation in Regular Higher Education Institutions*', which was formulated by the Chinese Ministry of Education in 2002 and revised in 2004. It served as the guideline for the undergraduate teaching evaluation of regular higher education institutions between 2003 and 2008.

Teaching staff

Teaching staff, or teachers (as often referred to in the interviews), in this book refers to those academic staff that work at regular higher education institutions and devote themselves to both teaching and research. I refer to them as teaching staff in order to highlight their role as educators. In addition, I have also selected my interview participants based on the criteria that they had to be teaching undergraduate course(s) at the time of my interview.

Student

Students in this book refer to undergraduate students who are pursuing bachelor's degrees. In the Chinese context, students refer specifically to undergraduate students at regular higher education institutions.

zhuanye, program

The Chinese term *zhuanye* (专业) mostly refers to a study program. In the Chinese higher education system, one program more or less corresponds to one discipline. In the interviews, the teaching staff and students used *zhuanye* to refer to a study program, discipline, major, or profession. I have adjusted the English term according to the context.

1

Introduction:
The rising concern over quality issues in higher education

'Quality' in higher education was not invented in recent decades; higher education institutions have always possessed mechanisms for assuring the quality of their work: for example, the qualification for admission and degree award, for promotion to professorship, appointment for an academic post, peer review of research and publications (Brennan & Shah, 2000, p. 2). However, public and government concern over quality in higher education has never been as conspicuous as it has been in recent decades.

The rising concern over quality is closely related to the change in higher education itself and its social context (Brennan & Shah, 2000; Dill, 2010; El-Khawas, 2007). The massive expansion of higher education across all continents has been one of the defining features of the late 20th and early 21st centuries. Since the 1980s, expansion (more students and institutions) in higher education has not only increased the costs and extended the numbers and types of people involved, which draws attention to issues of quality, it has also removed the prime traditional mechanisms for achieving quality – exclusiveness and selectivity (Brennan & Shah, 2000, pp. 20-21).

With expansion there is also a diversification dimension in the change in higher education where new types of institutions, programs, students and methods of delivery have joined in. And this raises the issue of engendering confidence in all the different parts, especially the private sector and the non-university sector which are relatively new things in some countries (Brennan & Shah, 2000, p. 21).

Expansion has also increased the cost of higher education everywhere while hardly anywhere has the funding level been able to keep pace with expansion (Brennan & Shah, 2000, p. 23). And this not only puts pressure on higher education institutions to seek alternative sources of funding and use funding efficiently, but also raises concerns over the accountability of public funding. Influenced by the New Public Management philosophy and the changing face of accountability in higher education, the states are taking a more utilitarian view of higher education in terms of seeking greater performance and productivity with limited investment, trying to acquire more value for resources (Alexander, 2000).

The New Public Management philosophy encourages governments to try to reform the public sector with a focus towards market-orientation, adopting private sector business principles to achieve greater efficiency. The emergence of formal quality assurance mechanisms such as evaluation, accreditation and audit in higher education by national agencies since the 1980s can be seen as a part of broader trends towards new forms of accountability in the public services and the professions. This trend was aptly characterized by Neave as the 'rise of the evaluative state' (Neave, 1988), or by Power (1997) as 'the audit society'. Recently, Neave (2009) summarized the evaluative state as retreating from direct control and adopting remote steering through assessing institutional output. Not long after starting the expansion of higher education in 1999, China also established a national quality assurance system – periodic evaluation of institutional undergraduate teaching.

Thus, the traditional internally enacted academic quality-keeping has been given an important external dimension – quality assurance, which requires higher education institutions to continuously demonstrate and improve performance, and which also provides new systems of rewards and sanctions.

However, recent policy trends in higher education, especially the complex impacts of quality assurance policies, have shown a need to further understand the quality issue in higher education, at least in China, as will be shown in Chapter 2. A review of current quality-related research has also shown an insufficient understanding of quality of higher education from the organizational and educational perspectives (see Chapter 3). This study attempts to contribute to this area. Specifically, this study is devoted to investigating the following research questions:

- How is quality of higher education perceived by the institution, teaching staff and students, respectively?
- What are the main concerns for the institution, teaching staff and students in their own pursuit of quality?

These questions attempt to find out the perceptions and concerns over quality of higher education from the points of view of the three core actors in higher education, i.e. the institution, teaching staff and students. This is to provide them with the opportunity to formulate their own views on quality. The assumption is that these core actors' views on quality form the essential parts of our understanding of the quality issue in higher education.

By addressing these research questions I am expecting to contribute to expanding our understanding of quality of higher education from organizational and educational perspectives, which would supplement the understanding from the currently most concerned policy perspective. Empirically, I have done my fieldwork in China mainly in the form of content analysis of institutional self-evaluation reports, teaching staff interviews and student focus groups. Thus, this study will also contribute by providing empirical data from the Chinese context. Based on my findings, I will reflect on the complexity of quality of higher education and possible implications for quality assurance.

The rest of the book is structured in the following way.

Chapter 2 provides a brief introduction to education in general and higher education in particular in China, which serves as the research context. It starts with a brief review of education in the Chinese context. This is followed by a general introduction to higher education in China, which consists of an overview, enrollment mechanism, institutional setting, key institution policy, etc. There then follows an outline of expansion and quality assurance in Chinese higher education.

Chapter 3 articulates the theoretical perspectives to approach the quality issue in higher education. With the facilitation of these perspectives the focus of this study, which looks into higher education from the organizational and educational perspectives, will be highlighted.

Chapter 4 addresses the methodology of this study, in which I will present my methodological assumptions, research design, methods of content analysis of institutional self-evaluation reports, teaching staff interviews, student focus groups and the analysis of data.

Chapters 5, 6 and 7 report respectively the findings from content analysis on institutional self-evaluation reports, teaching staff interview and student focus group. Three themes, i.e., student and learning, teaching staff and teaching, university and higher education, will be highlighted in these reports.

Chapter 8 first summarizes the findings from the previous three chapters and then interprets these findings in relation to their corresponding contexts.

Chapter 9 wraps up the whole report and reflects on the implications of the study.

2

Quality of higher education in the Chinese context

China has been selected as the context of this study. There were at least two considerations for this selection. First, in the English language academic world, China is an under-presented region in the issue of quality of higher education. For example, Harvey and Williams (2010) reviewed the articles in the journal *Quality in Higher Education* for 15 years from its creation in 1995 and showed that these articles covered discussion of quality issues in a lot of countries almost all over the world, but with no information about those in mainland China. Second, recent policies about innovation of higher education in China showed an urgent need to understand the current situation of quality assurance practice (as I will argue below).

This chapter aims at outlining the context of this study, which paves the way for the following presentation of research findings and discussion. It first gives a brief profile of education in the Chinese context. Then the focus will turn to higher education. The components of the Chinese higher education system will be outlined, followed by more information about the key university policy, higher education enrollment and institutional setting at regular higher education institutions. Finally, I will also look into one of the most dramatic changes in higher education in China, i.e. the expansion since 1999, and its associated quality assurance policies.

2.1 Education in the Chinese context

Education in China is a state-run system of public education, which is in the charge of the Ministry of Education. Usually it consists of six years of primary school, three years of junior secondary school, three years of high (senior secondary) school, and four years of university undergraduate education or three years of short-cycle college education (see Table 2.1). Following the undergraduate education there is also graduate education (2–3 years for a master's and three years for a PhD) at universities or research institutions. There are also more and more young children getting into the preschool institutions before they go to primary schools. The number of students and institutions in 2010 (Table 2.1) makes it clear that the Chinese educational system is one of the world's largest educational systems.

Table 2.1 Mainstream education in China[1]

Education	Years	Typical Age	Students in 2010	Institutions in 2010
Primary school	6	6–11	99,407,000	257,400
Junior secondary school	3	12–14	52,793,300	54,900
High school	3	15–17	24,273,400	14,058
University or college	3 or 4	18–22	22,317,900	2,358

Primary and junior secondary school education – nine years in total, is compulsory for all citizens. From the end of junior secondary school, there are selective entrance examinations for the following stage of education, i.e. there is the high school entrance examination right after junior secondary education, the higher education entrance examination right after high school education, and the postgraduate entrance examination after undergraduate education. Examinations are the most important mechanism that leads students into different tracks of education. These examinations can be seen as high-stakes tests. Usually the examination results determine whether students can go to the next stage of education or not, and which institutions they can go to. Students who secure better results in examinations are entitled to go to the key schools or universities with better resources and teaching staff. This better education is (or at least is believed to be) associated with better employment in China; and thus it is also believed to be the most important opportunity to expand one's life chances.

The Chinese culture also puts an emphasis on education. From Confucius comes the idea for the state to make people rich and educate them.[2] For individuals, Confucianism also assumes that everyone is

___

[1] This mainstream education refers to the education that is provided by the regular schools or institutions run by the state, which does not include the adult schools or institutions, occupational high schools, special education schools, non-state or private institutions, etc. The numbers of students and institutions in 2010 are from the Ministry of Education (2012a).

[2] See Zilu in the *Analects (Lunyu)*. In Chinese, 子适卫，冉有仆。子曰："庶矣哉！"冉有曰："既庶矣，又何加焉？"曰："富之。"曰："既富矣，又何加焉？"曰："教之。"

perfectible and educable; differences in intelligence do not inhibit one's educability (Lee, 1996, pp. 28–29). As a result of this emphasis on education, parents and teachers are known to attach great importance to education and achievement of their children and students (Leung, 1998). Education has been constructed as the most popular and legitimate approach to achieving personal success. Achievement in education has been connected with great prestige, and sometimes can in itself be a symbol of achievement in general.

2.2 A brief introduction to higher education in China

The following introduction starts with an overview of the whole higher education system in China, and then turns to the enrollment and institutional setting of undergraduate education at regular higher education institutions, which is the focus of this study. Finally, the key institution policy is also briefly outlined.

2.2.1 Overview of higher education in China

The higher education system in China is a large and complex one consisting of various institutions and programs. Generally speaking, there are four types of higher education providers in China (cf. Min, 2004):

- **Regular higher education institutions**, consist of universities and colleges that provide postgraduate programs with master's and doctor's degrees, undergraduate programs with bachelor's degrees, and short-cycle programs (two or three years, in Chinese *zhuanke*) with no degree (some of them also provide adult programs); they mainly enroll students from high school;
- **Research institutions**, provide postgraduate programs;
- **Adult higher education institutions**, consist of workers' colleges, peasants' colleges, institutes for administrative training, educational colleges, independent correspondence colleges, radio/TV institutions etc. that provide undergraduate programs and short-cycle non-degree programs;
- **Non-state/private higher education institutions**, provide undergraduate programs and short-cycle non-degree programs.

Table 2.2 below shows the size of different components in the Chinese higher education system in 2008. It can be seen from the table that mainstream undergraduate education at regular higher education institutions enrolls the most students among all kinds of education provided by the Chinese higher education system. Diploma from this kind of education is particularly highly valued in the job market and has the best social reputation among all the kinds of education of the same level. So this study sets its focus on this mainstream undergraduate education. I use the term 'undergraduate education' in this report to refer to this kind of education if there is no other specification.

Table 2.2 The Chinese higher education system in 2008[3]

Types of Institution	Number of Institutions	Number of Students
Institutions with postgraduate programs Of which:	796	1,283,046
Universities	479	1,230,945
Research institutions	317	52,101
Regular institutions with undergraduate programs and short-cycle programs Of which:	2,263	19,852,665
Institutions with bachelor's degree programs	1,079	11,032,160
Institutions with short-cycle non-degree programs	1,184	8,820,505
Adult higher education Of which:		5,482,949
Adult institutions	400	594,126
Adult programs run by regular institutions		4,888,823
Non-state/private institutions	866	4,013,010

Participating in the National Higher Education Entrance Examination is the primary way to get access to the undergraduate education at regular higher education institutions; and for more than 90% of students, it is the only way.

---

[3] The numbers of students and institutions in 2008 are from the website of the Ministry of Education, China (2009a, b, c, d, e).

2.2.2 Higher education enrollment and National Higher Education Entrance Examination

In China, the most important event for students to gain access to a university is the National Higher Education Entrance Examination, commonly known as the *Gaokao* in China. It is an academic examination held annually in mainland China. This examination is a prerequisite for entrance into almost all (regular) higher education institutions at the undergraduate level. It is usually taken by students in their last year of high school. The subjects to be tested are mainly those taught in high school. Three subjects are mandatory: Chinese, mathematics and a foreign language – usually English. The other subjects are three sciences, physics, chemistry and biology, and three humanities, history, geography and political education.[4] Applicants to science/engineering or art/humanities programs typically take one to three from the respective category. The actual requirement varies from province to province.

Performance in the examination is the main, and often the only, criterion for higher education admissions. Therefore, almost all students give most, if not all, of their energy in high school to studying and preparing for the exam. So do their high school teachers. Some students who haven't performed well, or think they haven't, will repeat the last year of high school life and make another attempt the following year.

In different places, students may be required to list their university or college preferences before or after the exam. The preferences are given in several tiers (including at least early admissions, key universities, regular universities and technical colleges), each of which can contain around four to six choices of institution and program. The institutions will admit students according to their performance in the exam and their preferences – those who prioritized a specific institution and perform well (get a higher score) in the exam are prioritized by the institution in its admission. Among students who have chosen the same institution, the ones with higher scores in the exam will have priority for being admitted to the program they prefer.

---

[4] This subject is partly a civic, introductory legal study, and partly ideology from the Communist Party of China.

2.2.3 The institutional setting of undergraduate study in China

For the students, university study is quite different from previous study in primary or secondary schools. In previous study, usually the contents are some fixed subjects/courses for years and the study is guided in detail by the teachers with a clear goal, i.e. to get as high a score as possible in examinations, especially the entrance examinations for high school or for higher education; and their lives are often taken good care of by their parents during this period. In university study, there are many more subjects, most of which are studied for only one semester and with different teachers. Often students can only interact with their teachers in classes. The course time is usually less than previous study periods in the primary and secondary schools. There is no longer one specific goal of study as they have in previous periods – a high score in exams may not be enough to get a job or go for further postgraduate study. During this university period they start to take care of themselves, to diversify their attention beyond study in schools and to be aware of complexity of the society and the challenges for survival (Zhang, 2005).

The undergraduate programs in Chinese universities normally last for four years, which are divided into eight semesters (except for some medical programs, which are usually five years divided into ten semesters). The teaching activity is mainly organized in the form of courses. Usually, a course consists of one or two lectures (90 minutes for each) every week and lasts for a whole semester (around 20 weeks). The teacher mainly acts as a lecturer. At the end of the semester there is an examination for each course, which mostly takes the form of a paper-and-pencil test (a few courses may also assess students through a paper they submit). Some courses in science and engineering programs may have some of the course time spent in the laboratories where the students perform experiments. A few teachers may design some projects as assignments within the course. Usually, one to four credits are assigned to each course based mainly on the course time each week; the more course time each week, the more credits are assigned to it.

Students can select courses for themselves before the start of each semester; however, some courses (the majority) are required and they have to choose them. Most of the time and on average they have about seven courses each semester. They have to collect about 150–180 credits and finish a thesis or design in the final semester in order to get their diploma. In the curriculum, there is usually also a requirement for internship, which takes place in the third year or the beginning of the fourth year in most cases.

Besides course teachers, who are responsible for and in charge of academic affairs for the students, there is another type of staff that also has the title of teacher – the form teacher or student mentor who is in charge of student affairs (daily management of students, issues other than academic ones) and helps them with adjustment to university life, activity organization, psychological facilitation, secretarial work for the students, career consulting, etc.

University students can acquire most (almost all) of the necessities for life and study within the campus – besides libraries and other teaching infrastructure, there are usually canteens, hospitals, dormitories and sports facilities which are all part of the university, and often there can be shops, restaurants, post offices, banks and buses within the university campus. There are usually clear geographical demarcations between the university campus and the outside world, often in the form of boundary walls. Almost all the students live in the university dormitories within the campus. Most students usually spend most of their time within the campus during their study period (normally four years for undergraduate students, around age 18 to age 22). It would be no problem for some students to live a whole semester within the campus without going out if they wished.

2.2.4 The key institution policy in China – 'Project 211' and 'Project 985'

As we can see from Tables 2.1 and 2.2, there are more than 2,000 higher education institutions in China. Some of them are more prioritized and more intensively sponsored by the government than others, which is most conspicuous in the key university policy, especially 'Project 211' and 'Project 985'.

'Project 211' is a project of national key universities and colleges in the 21st century initiated by the Chinese government in 1995, which aims to specially advance around 100 key institutions (Ministry of Education, 2008). Higher education institutions in this project are prioritized for being sponsored by national funds. The figures of 21 and 1 within 211 are from the abbreviation of the 21st century and approximately 100 universities respectively.

'Project 985' is a project to promote some key universities (around 40 so far) to become world-class level (Ministry of Education, 2011). It was announced by President Jiang Zemin on the 100th anniversary of Peking University on May 4, 1998 (the code is named after the date, year 98 month 5). The objective of the project is to promote the development of Chinese universities so as to raise their influence and reputation in the world. Large amounts of government funding have been allocated to universities admitted into this project. The '985 institutions' are also included in 'Project 211'.

Since these institutions are more intensively supported by the government in terms of more resources and other privileges, being included in 'Project 211' or 'Project 985' is often in itself a symbol of prestige or good quality.

2.3 Expansion and quality assurance in Chinese higher education

The concern over the quality issue in higher education is closely associated with its expansion in China.

2.3.1 Expansion of higher education in China

After the ten-year Culture Revolution, where there was hardly any development, China started new state policies aimed at speeding up economic development in the late 1970s. These policies mainly consisted of economic reform and opening up to the outside world, which initiated the long process of transition from the centrally planned economy to a market economy. With the establishment of a market economy since 1992, education was considered the strategic foundation for economic success in terms of well-educated manpower, especially high-level specialized personnel (Min, 2004). The fast-growing market economy, the rapid development of science and technology and the increase in individual income levels and living standards stimulated increasing demands for higher education (Min, 2002). The development of the economy increased the demand for highly educated workers; individuals and families now wanted to invest in higher education as a means to secure both a higher income and status in society, and they could afford to do so (Li, Morgan, & Ding, 2008).

At the end of 1998, facing the impacts of the Asian economic crisis and slow economic growth, the government adopted economist Min Tang's advice (Tang & Zuo, 2004) about expanding higher education to stimulate economic development, and launched the higher education expansion policy in order to 'increase domestic demand, stimulate consumption, and alleviate employment pressure' (Feng & Li, 2009).

Figure 2.1 shows clearly the effect of the expansion policy. There is a smooth increase of students in regular higher education institutions in the 20 years between 1978 and 1998. With the launch of the expansion policy, such as the '*Action Scheme for Invigorating Education Towards the 21st Century*' at the end of 1998, student numbers increased almost four times in just nine years. The gross enrollment rate increased from 9.8% in 1998 to 23.3% in 2008 (Ministry of Education, 2000, 2009a). Table 2.1 above also shows that there were relatively approximate numbers of high school students and students in undergraduate or short-cycle higher education in 2010.

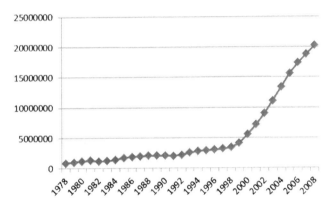

Figure 2.1 The increase of students in regular institutions[5]

This expansion offers opportunities for a much larger population to go through higher education. However, a number of issues have appeared with such a rapid expansion, among which are insufficient teaching resources, employment difficulties for graduates and doubts about quality (Gan & Deng, 2008; Li, 2000; Sun, 2001; Yang & Chen, 2002; Zhang, 2006).

The rapid increase of students brought great challenges to higher education institutions. The resources, facilities, and staff, especially teaching staff, could not expand at the same speed or in proportion with the increase in students. Thus, there appeared many big classes of more than 100 students. In this situation, the quality issue of higher education attracted more and more attention from both the government and the public.

---

[5] Postgraduates not included. See Zou, Du, & Rasmussen (2012).

In 2004, the Ministry of Education released the *'Action Plan of Education Innovation 2003–2007'*, which initiated to establish a mechanism for quality assurance in higher education (Ministry of Education, 2004b). On May 10, 2006, at a meeting of the standing committee of the state council, Premier Wen Jiabao pointed out that the expansion of higher education should be controlled, and the focus on higher education should be shifted to quality enhancement (Zhou, 2007). In 2007, the central government endorsed the *'Outline of the 11th Five-year Plan for National Educational Development'* drafted by the Ministry of Education, which formally and explicitly announced the focus on quality enhancement in higher education. In July 2010, the *'Outline of the National Plan for Medium and Long-term Education Reform and Development (2010-2020)'* was issued by the Chinese central government, highlighting quality improvement as the central task in higher education. And recently the Ministry of Education (2012b) released the *'Memorandum on Comprehensively Enhancing Higher Education Quality'*, which listed 30 specific measures to improve the quality of higher education in China.

These policies formed the national policy discourse on higher education development. They showed an increasing importance of the quality issue in higher education. To some extent they also reflected the challenges that came out with expansion. Among those policies, the most notable procedure is the establishment of a national system of quality assurance – institutional evaluation of undergraduate teaching.

2.3.2 Evaluation of higher education institutions – A Chinese approach to quality assurance

After a few years of expansion, the main concern over higher education in China has shifted from the previous insufficient opportunities to its quality. Since 2002, China has begun to formally respond to the challenges brought by expansion in the form of developing a nationwide formal evaluation system. The *'Plan for Undergraduate Teaching Evaluation in Regular Higher Education Institutions'* (referred to as the **Evaluation Plan** hereafter) was formulated by the Ministry of Education in the same year. And the first round of undergraduate teaching evaluations was guided by a revised version of this plan (Ministry of Education, 2004a). In addition, *'Action Plan of Education Innovation 2003–2007'* further confirmed that all higher education institutions were required to undergo a quality evaluation every five years (Ministry of Education, 2004b). The Higher Education Evaluation Center (HEEC) was established in 2004 to serve as the national coordinating body for the evaluation. Since then, the formal higher education evaluation system has gradually taken shape.

In 2007, the Ministry of Education and Ministry of Finance co-launched the *'Undergraduate Teaching Quality and Innovation Project'* (known as the *'Quality Project'*) aimed at quality assurance and improvement. In addition to confirming the practice of the above-mentioned periodic teaching evaluation, this policy also stipulated steps for (1) program restructuring and program accreditation, (2) optimizing the curriculum through the *National Excellent Course Program*, improving textbooks, and sharing teaching resources, (3) innovation of the educational models and approaches, (4) developing teaching teams and high-level faculties, (5) building teaching databases to publish teaching-related information, (6) supporting institutions in western China through faculty exchange programs (Ministry of Education & Ministry of Finance, 2007).

The first round of institutional evaluation of undergraduate teaching was conducted between 2003 and 2008. A total of 589 institutions were evaluated in this round. According to the Evaluation Plan, the goals are

'to enhance the state macro management and steering of the teaching work in higher education institutions; to emphasize and support teaching in institutions from education administrations of all levels; and to help institutions implement education policies by further specifying their guiding ideas for running the institution, improving the institution's conditions, enhancing the teaching infrastructure and administration, promoting teaching reform, and generally promoting teaching quality and effectiveness' (Ministry of Education, 2004a).

The evaluation starts with institutional self-evaluation where a self-evaluation report is produced and submitted to the ministry. Then a group of academic peers conduct a site visit and review the institutional activities, which is partly based on the self-evaluation report. After the peer review, the feedback is given to the institution, and the evaluation results, in the form of a grade of *Excellent, Good, Acceptable,* or *Not Acceptable*, is released publicly. Finally, the institutions will produce an action plan for improvement.

The primary indicators of the evaluation are (Ministry of Education, 2004a):

- Guiding ideas for running the university
- Teaching staff

- Teaching facilities and their usage
- Program construction and teaching innovation
- Teaching management
- Learning atmosphere
- Teaching outcomes
- Special characteristics

2.3.3 Reflections on quality assurance of higher education in China

As can be seen from the above brief introduction, the expansion of higher education in China was, to a large extent, stimulated by economic considerations, and the expansion shifted the main concern over higher education from opportunity to quality. Then came the government-initiated institutional evaluation as the main mechanism for quality assurance.

The institutional evaluation seems to have had complex impacts (Chen, 2009; Lee, Huang, & Zhong, 2012; Zhong, Zhou, Liu, & Wei, 2009): it made central government play the role of a regulator and a coordinator rather than that of a direct administrator; it also stimulated the institutions to develop clear development orientation plans, to highlight the importance of teaching among the institutional activities, and to make efforts at educational reform and enhancing administrative efficacy. However, there are quite a few deficiencies to this institutional evaluation, such as using the same indicators to evaluate different types of institutions, lack of differentiation in results (an overwhelming majority are excellent), exerting too much pressure on the institutions, formalism leading to a waste of resources and professional misbehavior, disturbing regular teaching activities, etc. (Chen, 2009; Lee, Huang, & Zhong, 2012; Zhong *et al.*, 2009). According to Gao *et al.* (2009), the administrative staff is more positive about the evaluation than academic staff; the effects are more on administration than students' study.

In those quality-related policies and practice guidelines such as the *Evaluation Plan* or the *quality project*, there are no explicit formulations on what quality means. It seems that these policies have assumed that all the actors involved in higher education have shared an unambiguous and uncontroversial understanding of quality. The complex impacts that appeared in the institutional evaluation showed a need for further understanding and reflection on quality in higher education, especially those on the frontline people – the teaching staff and students. All the quality assurance procedures are initiated and administered from above, i.e. the government. To take further action, the government urgently needs to know more about the quality of higher education and how higher education works from the perspective of those on the front line.

3

Perspectives to study quality of higher education

To gain further understanding of the quality issue in higher education, it is necessary to construct some theoretical tools and put the investigation into the current research context. This chapter mainly aims at this goal. It begins with a brief review of quality assurance in higher education. Then the three perspectives of policy, organization and education are sketched as theoretical tools to approach the quality issue in higher education. Then these perspectives are used to map the current research on the quality issues in higher education, which paves the way to highlight this study's focuses on the organizational and education perspectives, and also the possible ways to answer the research questions in this study.

3.1 Quality assurance in higher education

Unlike industry, where quality management issues arose from within, the quality imperative for specialized quality management in higher education mainly came from the market and from government (e.g. Idrus, 1996; Salter & Tapper, 2000). In practice, not only is the concept of quality assurance in higher education inspired by industry (Ellis, 1993), but also the language and tools of industry-born quality models are imported by higher education (Houston, 2008), for example, audit, accreditation and benchmarking.

Although in practice audit, accreditation and evaluation can be seen as different approaches that are adopted for quality assurance in different social cultural contexts, van Vught and Westerheijden (1993; 1994) found there are common elements among them and offered a general model for quality assurance: (1) a national coordinating body, (2) institutional self-evaluation, (3) external evaluation by academic peers; and (4) published reports. Brennan and Shah (2000, p. 52) elaborated the model as follows. The national coordinating body is responsible for coordinating and setting out the procedures and methods to be used by institutions of higher education for the assurance of quality. Based on the procedures and methods set out by the national coordinating body, institutions should undertake regular self-evaluation and report to the coordinating body on a regular basis. This forms the basis of an external peer evaluation, which usually takes the form of a site visit. The main purpose of the report is to make recommendations to institutions in order to help them improve the quality of their teaching and research. But the language and tools of industry-born quality models seem to be significantly questionable in higher education (Harvey, 1995; Houston, 2008).

Plainly adopting quality approaches from outside the university largely ignores the views of academics, students and others affected within the university (Ulrich, 2001). In general, what these approaches of quality assurance share in common is regarding higher education as a total system, in which students enter as inputs, are processed, and emerge as outputs; however, important concerns are raised: for example, the learning experience of students is neglected, and limited insight into the educational processes is offered, which are at the heart of higher education enterprise (Barnett, 1992, p. 20). These approaches to quality and the practice of quality assurance are dominated by the input/output methodology, where the viewpoint of the outsider and the extrinsic justification are highly favored, while the process, the viewpoint of the insider and the intrinsic justification remain under-presented. There is no framework in which the institutions can talk about the quality of higher education educationally (Barnett, 1990: 3-4). From this point of view, little can be learned concerning how the educational processes can be improved. Therefore, they could only play very limited functional role in quality assurance and improvement.

From the above preliminary review, we can see that quality assurance in higher education is, to a large extent, externally-originated and mainly initiated by the government. Internally, higher education institutions and academics seem to have quite different views on the quality issue of higher education. So in order to understand the complex issue of quality, a clarification of the concept and the perspectives to approach it is probably helpful.

3.2 Quality of higher education – the concept

The term 'quality' in our everyday use is, like 'good', a signal that the user is picking out some element that he or she wishes to endorse; it simply betokens an expression of positive inner feeling towards the feature being picked out (Barnett, 1992, pp. 2–3). Or to use dictionary terms, quality mainly refers to (1) the attribute, property, character, nature, feature of something or somebody, or (2) being of good nature, positive characteristic, the degree of excellence based on the attributes or properties of something (Merriam-Webster Dictionary, n.d.; Oxford English Dictionary, n.d.).Thus, quality essentially involves a normative judgment about values where value as Bauman (2008, p. 38) suggests, 'by its nature, is always outstanding, always somewhat ahead of what is: nothing that is-already-here-and-now can accommodate value in full'. People may pursue different values at the same time which may cause some kind of tension.

In business, engineering and manufacturing it has a pragmatic interpretation as the non-inferiority or superiority of something. It may be understood differently by different people. For example, consumers may focus on what a product/service is like by comparing alternatives in the marketplace. Producers tend to measure the conformance quality, or degree to which the product/service was produced correctly, free of deficiencies. From a management point of view, quality has been defined variously as value, providing good value for costs (Abbott, 1955; Feigenbaum, 1951), as conformance to specifications (Gilmore, 1974; Levitt, 1972), as conformance to requirements (Crosby, 1979), as fitness for use (Juran & Gryna, 1988), and as meeting and/or exceeding customers' expectations (Gronroos, 1983; Parasuraman, Zeithaml, & Berry, 1985), etc.

Although in practice quality often has multiple definitions and has been used to describe a variety of phenomena, we can still identify something in common – quality can be understood generally both as the attributes/properties (no deficiencies, durability, brightness, etc.) of something and as the degree of these attributes/properties in relation to certain criteria (this can be costs, specifications, standards, requirements, needs, use or expectations, etc.).

Similar to the quality concept in general, the concept of quality of higher education is also a complex one. Different views may be adopted by various stakeholders, i.e. students, academics, institutional administrators, employers, governments and professional organizations. The concept of quality of higher education is often contested for at least two reasons: first, different stakeholders have different concepts of higher education and they focus on different attributes/properties of higher education; second, they give different references or criteria to higher education when talking about quality. The above brief discussion also offers us one way to approach quality – to find out attributes and properties that people concern, and the criteria given for judging them. To figure out the attributes and criteria, a differentiation and clarification of perspectives is often necessary.

3.3 Perspectives to approach quality of higher education

Inspired by House and McQuillan's (2005) three-perspective (technological, political and cultural) analysis of successful school reform, there can also be different perspectives from which to look into the quality issue of higher education; and some of them have already been briefly touched on above. If we limit the concern to the three main directly involved actors in higher education, i.e. the government (as a major sponsor and the regulator), the higher education institution (as the implementer and local decision-maker), and the academic/teaching staff and the students (as the front-line people), there appear to be three broad conceptual perspectives that can be taken: the policy (or government) perspective, the organizational (or institutional) perspective, and the educational (or teaching and learning) perspective. These three perspectives are also the most distinguishable ones in the existing literature. However, it should be noted that the perspectives are analytical constructs highlighting the way of thinking from a certain standpoint, which may not exist in its pure form in reality.

3.3.1 The policy perspective

The policy perspective is mainly taken by the ideal government, which takes the common good as its only concern without any self-interest. It tries to promote and facilitate higher education institutions to complete their tasks properly, effectively and efficiently, and to hold the higher education institutions accountable for its utilizing public financial support and other public resources. From this perspective, the discussion on quality mainly concerns the extent to which higher education institutions effectively and efficiently pursue their goals and social functions. In policy analysis rationale, the values pursued may be efficiency, human dignity, distributional equity, economic opportunity, or political participation (Weimer & Vining, 2005, p. 55).

The keywords of the policy perspective are the role of higher education in society, accountability, efficiency of using public resources, etc. The questions asked from this perspective are: What is the role of higher education in society? How can higher education institutions be promoted and facilitated to realize their roles, and realize them efficiently? How can the standards of higher education be ensured?

With the rise of the New Public Management philosophy and accountability discourse in the recent administrative reforms, policymakers have introduced a range of private sector management styles and instruments into the public sector, including contract management, customer-oriented management, and the use of performance indicators and benchmarks to evaluate and compare the effectiveness and efficiency of public agencies (Bovens, 2005; Pollitt & Bouckaert, 2004; Riccucci, 2001). Most of the instruments require extensive auditing to be effective.

Being part of this trend (or at least significantly influenced by this trend), different approaches to quality assurance – for example, institutional audit, evaluation, accreditation and benchmarking – have been adopted in the administration of higher education worldwide. With China looking into the Western world for inspiration about administrative reform (Christensen, Lisheng, & Painter, 2008), this trend also has a significant influence on the Chinese public sector administration, including higher education.

Research on quality of higher education from this perspective looks into quality-related policies in higher education to see how these policies have addressed the common good and how they can be improved; for example, what has been constructed as quality in the policy discourse and various approaches for quality assurance such as accreditation, evaluation, audit, institutional review, benchmarking, etc. and whether these constructions and related policy tools are appropriate.

3.3.2 The organizational perspective

The point of departure of the organizational perspective is to regard higher education institutions as organizations interacting with the environment (their host society). It looks into how organizations can survive and prosper in their interactions with the environment. All organizations must pursue or support 'maintenance' goals in addition to their output goals (Gross, 1968; Perrow, 1970, p. 135). An effective organization must also respond to the demands of its environment according to its dependence on the various components of the environment (Pfeffer & Salancik, 2003, p. 84). While achieving their educational goals, higher education institutions are simultaneously struggling in raising resources from the environment for survival and organizational prosperity.

The keywords of the organizational perspective are social recognition, fundraising, reputation, legitimacy to acquire resources, etc. The question asked from this perspective is: How can higher education institutions survive and prosper in their host society? Or to put it more specifically, how can the institutions respond to the external requirements of quality assurance to achieve its organizational survival and prosperity?

In many countries, including China, the government is the main resource provider for the higher education institutions. Thus they are significantly dependent on the government. And it is important for them to the acquire government's recognition of effectiveness and efficiency in order to be able to apply for more resources. However, it is often more difficult for higher education institutions to evidence their effectiveness and efficiency than other organizations. Companies can use the profitability criterion in internal decision-making, e.g. in closing down or expanding activities, while higher education institutions are confronted with more complex demands and have to use more complex criteria (Zou *et al*, 2012). This situation has been described by some scholars as organized anarchy in order to characterize their dependence on people with specialized skills and professional autonomy (cf. Birnbaum, 1988; Clark, 1983; Cohen and March, 1986). Therefore, quality assurance requirements probably bring significant challenges to the institutions.

Research on quality of higher education from this perspective looks into how higher education institutions as organizations respond to quality assurance. For example, what has been constructed as quality in the institutional discourse and how higher education institutions demonstrate their quality to the external environment, especially to the government.

3.3.3 The educational perspective

The educational perspective focuses on the development and growth of individual students, and how to facilitate this process. The questions asked from this perspective are: What is (good) learning? How does it happen? What can be done to facilitate learning, and how? It looks into what teachers and students perceive as quality in higher education and tries to extract the teaching and learning perspective of quality from their perceptions.

There are different theories trying to address the contents of learning and how learning happens. Based on a review of some major learning theories, Illeris (2003, 2007, 2009) proposed a synthesis model to understand the processes and dimensions of learning, which paved an illuminative perspective to understand the quality of higher education. He pointed out that all learning includes two different processes: an interactive process between the individual and the environment, and internal mental acquisition and processing through which impulses from the interaction are integrated with the results of prior learning. Acquisition always includes content and incentive, therefore, learning consists of three dimensions: content, incentive and interaction. The content dimension typically concerns knowledge, understanding and skills, through which we strengthen our functionality, i.e. our ability to function appropriately in the given environment. The incentive dimension comprises motivation, emotion and volition, through which we seek to maintain mental and bodily balance and at the same time develop our sensitivity. The interaction dimension includes action, communication and cooperation, through which we seek to achieve social and societal integration that we find acceptable, and at the same time develop our sociality.

Illeris's synthesis can be a framework to understand learning in general. When it comes to the quality issue of higher education, the question is how the arrangements of the institutions have assisted students in these processes and dimensions. There is also something specific for higher education. For example, Barnett (1990, p. 202) argued that an educational process can be termed higher education when the student is carried to levels of reasoning which make critical reflection on his or her experiences possible, whether consisting of propositional knowledge or of knowledge through action. These levels of reasoning are 'higher', because they enable the student to take an above view of what has been learned. Simply, 'higher education' resides in the higher-order state of mind. Brubacher (1982, pp. 2–3) also proposed that the principal difference between higher and lower education is that higher education is capped by higher learning. Here, 'higher', as used by Brubacher, not only refers to the upper reaches of the educational system, but is more concerned with the highly sophisticated knowledge involved, which 'is either on the very frontier between what is known and unknown or, if known, is so esoteric and arcane that it escapes the grasp of the average person's intelligence'.

The educational perspective on quality in higher education provides a vocabulary which takes seriously the educator's aims towards the individual students. It is a vocabulary in which the arrangements in higher education and their effects on students are central (Barnett, 1992, pp. 6–7). The key questions asked are: What is it to educate in higher education?

What is offered to students in higher education? Based on an understanding of these questions, it asks: what is it to assess and to improve the quality of higher education? It is an approach that takes its bearings from a sense of the educational mission of higher education. Since teaching is the core activity in the educational provision, it is also the main focus of this perspective.

3.3.4 The three perspectives and the historical development of higher education

In an ideal situation, the government provides higher education institutions with resources to complete the educational missions; the institutions direct their staff to devote themselves to the educational enterprise to achieve its organizational effectiveness and thus gain the legitimacy for resources. This would be a smooth and conflict-free system. However, this is definitely not the situation in reality.

Historically, the recognizable institution of the university – a school of higher learning combining teaching and scholarship – rose up in the form of corporate autonomy and academic freedom in medieval Europe (Perkin, 2007). Rooted in the university's historical development, the academic spirit and educational perspective have taken their form relatively independently. In our contemporary world, universities in most countries have become more and more dependent on the state for resources. The government has been trying to influence universities through its policy tools while providing them with resources. For example, the OECD countries have come to hold ambitious goals for higher education, viewing it both as a means to foster economic growth and as a principal instrument for the fostering of social cohesion. Or to put it more specifically, these countries are promoting higher education to contribute to creating a highly skilled workforce and research that underpins a knowledge-based economy, and at the same time to widely disperse the benefits of economic growth. To make the institutions more accountable for the accomplishment of these public purposes, governments have employed governance tools such as performance or output control, performance reporting, performance contracts, etc. (OECD, 2006).

Thus, there are at least two major streams with different priorities on the agenda that are trying to shape the organization of universities – state policy and academics. This is where the three perspectives meet and tensions, if not conflicts, probably exist here. For example, as already indicated above, the government, influenced by the New Public

Management philosophy, started to demand accountability in terms of effectiveness and efficiency, while it is difficult for higher education institutions to demonstrate their effectiveness and efficiency.

3.4 Mapping current research with the perspectives from the policy, organization and education

With, or as part of, the rising concern over the quality issue, more and more research has looked into it; and there are even specialized academic journals on quality such as *Quality Assurance in Education* (launched in 1993) and *Quality in Higher Education* (launched in 1995). But the three perspectives to look into the quality issue in higher education sketched above are not evenly covered in the current research.

3.4.1 The highlighted focus on policy

Parallel to the external origination of the quality assurance concerns over higher education quality, most of the researches on quality and quality assurance in higher education are from the policy perspective (cf. Harvey & Williams, 2010). The frequently discussed topics are different approaches to quality assurance in various national or regional contexts (Andersen, Dahler-Larsen, & Pedersen, 2009; Bornmann, Mittag & Daniel, 2006; Dill, 2000; García-Aracil & Palomares-Montero, 2010; Kettunen, 2008; Meade & Woodhouse, 2000; Quinn & Boughey, 2009; Schwarz & Westerheijden, 2004; van Kemenade & Hardjono, 2010; Woodhouse, 2003), mutual recognition of qualifications and diplomas in the case of cross-border student and staff mobility (Heusser, 2006; van Damme, 2001; Van der Wende & Westerheijden, 2001; Vroeijenstijn, 1999;), the impact of quality assurance (Nilsson & Wahlen, 2000; Stensaker, Langfeldta, Harvey, Huisman, & Westerheijdend, 2011; Volkwein, Lattuca, Harper, & Domingo, 2007).

While it is generally recognized that institutions must be held accountable for the quality of their activities, there is an abundance of different interpretations of quality (Kitagawa, 2003). Based on 29 case studies of quality assessment practice in different countries, Brennan and Shah (2000: 14-15) identified four major values in the current practice of assessing the quality of higher education: the academic, the managerial, the pedagogic and employment-focused. The academic values are the traditional ones which focus upon the subject field and their criteria of quality stem from the characteristics of the subject. The managerial values are associated with the institutional focus of assessment, with a concern about procedures and structures, with an assumption that quality can be produced by 'good management'. The pedagogic values focus on people, on their teaching skills and classroom practice. The employment-focused values emphasize graduate output characteristics, standards and learning outcomes. It is an approach which takes account of 'customer' requirements where the customers are frequently regarded as the employers of graduates.

Closely related to the values at work in quality assurance, Barnett (1992, pp. 18–20) identified four prevailing concepts of higher education underlying contemporary approaches to quality: (1) higher education as the production of qualified manpower; (2) higher education as training for a research career (which corresponds to the above-mentioned academic value); (3) higher education as the efficient management of teaching provision (corresponds to the above managerial value); (4) higher education as a matter of extending life chances (roughly corresponds to the above employment-focused value). Barnett also pointed out that all of these four conceptions regard higher education as a total system, in which students enter as inputs, are processed, and emerge as outputs, and in which the educational experience of students is neglected. Thus, there is a lack of voice from education in these approaches to quality in higher education.

After reviewing the indicator systems for evaluating universities applied in the OECD countries, García-Aracil and Palomares-Montero (2010) demonstrated the difficulty in establishing classification criteria for existing indicators, on which there is currently no consensus. This is mainly because of the multiple objectives of higher education and the variety of principals (actors) and stakeholders involved.

3.4.2   Mismatch of policy rhetoric and impacts

Having reviewed total quality management, performance indicators and external quality monitoring in higher education, Law (2010) criticized all these quality assurance approaches as being mainly driven by demands to satisfy external agendas (e.g. to enforce institutional accountability or compliance) instead of academic considerations (e.g. to facilitate student learning). As a result, a mismatch between the policy rhetoric and reality of educational quality has become a common experience for most practitioners. Newton (2000, 2002) and Cartwright (2007) also found that the academic staff perceived an implementation gap or mismatch between the intentions underpinning quality policy and the actual outcomes.

In addition, different groups of people in higher education appear to have different perceptions of the impacts of quality assurance (Stensaker et al., 2011). According to Stensaker et al., the leadership and administration members perceived more positive effects of evaluations and accreditations than the academic staff and students. The academic staff only saw clear positive effects in prioritizing research and improved infrastructure in research. The students identified the fewest clearly positive results from evaluations; and a large part of the evaluation effects they didn't know. So quality assurance seems to be more about bureaucracy, organization and regulation than about issues that are central in the minds of the academic staff and students. A historical study on the British system of quality monitoring by Harvey (2005) also concluded that the system failed to engage with transformative learning and teaching; he pointed out that quality monitoring focused on processes and systems rather than engaging with the learning experience of students. The historical analysis reveals how quality evaluations were guided by political pragmatism as much as by rational evaluation.

Resonating to Harvey's research in the UK, Vieira (2002) also found that teachers and students in Portugal shared an idealized conception of pedagogic quality that was in accordance with a view of pedagogy as a process of emancipatory transformation, but the reality was felt to be far from expectations and ideals. There were also different perceptions about pedagogic quality: teachers were more concerned about transparency and coherence than students, and students felt the absence of reflectivity and self-direction; teachers tended to locate some main problems in students and institutional factors, while students seemed to stress the institutional factors, and were not as positive about teachers as teachers perceived themselves.

After articulating the 'implementation gap', Newton (2000) argued that if academics are to remain pivotal in efforts to improve the quality of teaching and learning, then more attention needs to be paid, by institutions and external quality bodies, to the importance of the conditions and context of academics' work. Otherwise, quality monitoring is liable to be invested with a 'beast-like' presence needing to be 'fed' with ritualistic practices by academics seeking to meet accountability requirements.

Although not very much research on quality of higher education has involved academic staff or students, it is clear from the above-mentioned research that academic staff and students have different views on quality assurance from that of policy statements and administrative members. The academic staffs' and students' own points of view on quality are still unclear and need to be explored in order to form a more inclusive understanding of the quality issue in higher education.

3.4.3 The under-researched organizational perspective and educational perspective

Compared to the policy perspective, there are far fewer studies trying to approach quality of higher education from the organizational perspective and from the educational perspective. There are a few studies trying to outline the strategic responses or initiatives adopted or possibly adopted by higher education institutions in the contemporary world (Clark, 1998; Gumport, 2000; Kezar, 2001; Shattock, 2010). Only a few studies directly addressed the quality issue from the organizational perspective in an effort to outline the organizational impacts of quality assurance (Liefner, 2003; Nilsson & Wahlen, 2000). There is insufficient articulation on quality of higher education from the organizational perspective, which is often necessary for higher education institutions to gain a voice in quality-related issues, and for the government to gain more understanding of how the institutions work.

Regarding the educational perspective on quality, although some scholars emphasized its importance and even started theoretical formulations (Barnett, 1992; Keeling & Hersh, 2011), there is still a lack of empirical investigation and evidence about what quality means for the actors involved. While academics and students have been asked how they perceived the impacts of the quality assurance policies, as mentioned above, there is still not much formulation of quality in higher education from their own points of view.

3.4.4 The significance of involving more perspectives on quality in higher education

In addition to the 'outside' approaches to quality, which provide justifications and accountability for higher education, it is necessary to go one step further – to complement these approaches with a look inside the system, to see what is inside the 'black box' of higher education, how it operates (especially the core activities – teaching and learning), and what can be done to improve. Questions of cost, efficiency, completion and access, important as they are, are relevant only if students learn (Keeling &

Hersh, 2011). Authentic quality improvement is more likely to result from approaches to systemic intervention that encourages exploration of questions of purpose and of the meaning of improvement in context (Houston, 2008).

From the above-discussed mismatch between quality assurance policy rhetoric and impacts, we can also see there is a need to employ a more inclusive view on the quality of higher education. As indicated in Chapter 2, the Chinese practice of institutional evaluation also demonstrates a need to further understand the quality issue. The educational perspective and organizational perspective are of essential importance for this more inclusive view since these perspectives are to illuminate how the core actors in higher education (i.e. the institution, the teaching staff and the students) work, and how quality assurance policy would probably impact them. More understanding on how the institution, the teaching staff and the students perceive quality and what their concerns are in the pursuit of quality would certainly shed light on new higher education policies.

3.5 Researching actors' perceptions and concerns about quality of higher education

Quality of higher education is not something separate from the actors involved in higher education, but rather resides in their perceptions, activities, experience and relations among each other. Thus, to know more about quality of higher education essentially involves knowing about people, especially their perceptions and concerns about various elements in higher education.

Based on the conceptual analysis in Section 3.3, the meaning of quality can be summarized as the attributes or goodness of something and the criteria used to judge those attributes. Thus, these actors' perception of quality of higher education may refer to how they make sense of higher education, especially those aspects and characteristics of higher education that they pay attention to and put emphasis on. Since higher education is not a tangible thing, further substantiation of the actors' perceptions of quality of higher education can be done in terms of their perceptions of the major elements of higher education – the institution and the higher education that the institution provides, teaching staff and their teaching, students and their learning. If we limit our investigation to the three core actors mentioned above, i.e. the institution, teaching staff and student, a

3 Perspectives to study quality of higher education

conceptual tool (Figure 3.1) can be developed to investigate quality of higher education from the organizational and educational perspectives.

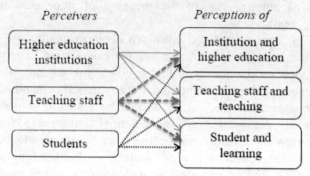

Figure 3.1 Investigation guide for perceptions of quality of higher education

Therefore, the investigation on quality of higher education is substantiated in the form of an investigation on core actors' perception of the major elements in higher education. This also enables us to cross-check their ideas about each other – to put their perceptions and expectations of each other side by side, and then to formulate an interconnected and more inclusive picture of the quality issue in higher education. In addition to that, this offers the actors something specific and relevant to talk about as well.

The actors' perceptions of quality of higher education often, if not always, form an important basis for their concerns over higher education, which are more closely associated with their actions that significantly influence the attributes of higher education.

The actors' concerns refer to the issues they give special attention and energy to in their work or study. These issues are the ones all actors struggle with and make an effort to achieve in their pursuit of quality of higher education. The concerns can be unfulfilled expectations or goals the actors are endeavoring to achieve; they can also be tensions and contradictions between inconsistent values or goals simultaneously pursued by the actors. For example, the institutions might be trying to play up to the government advocating contributing more to economic growth, so they can acquire more resource support from the government; and at the same time the institutions are also devoted to intellectual and cultural values, which may not be totally consistent with the economic pursuit. Therefore, there might be a tension between these different pursuits for the institution. For the teaching staff there might be a tension between teaching and research, which compete with each other for time and energy allocation.

3.6 Researching actors' perceptions and concerns 33

As lives lived and lives told are closely interconnected and interdependent (Bauman, 2001, p. 7), people's concerns are also reflected in their perceptions. Thus, besides directly addressing questions about concerns, proper interpretation of these higher education actors' perceptions in relation to their corresponding contexts would also significantly contribute to the investigation into their concerns. To be specific, the investigation of each actor's perceptions of quality of higher education in relation to the institution and higher education, teaching staff and teaching, students and learning would be an important way to investigate their concerns in their own pursuit of quality of higher education.

3.6 The focus of this project

Based on the above review of literature, this project attempts to explore quality of higher education from the organizational and educational perspectives, which involve the three core actors in higher education, i.e. the institution, the teaching staff and the students. The research questions addressed are:

- How is quality of higher education perceived by the institution, teaching staff and students, respectively?
- What are the main concerns for the institution, teaching staff and students in their own pursuit of quality?

Empirically, it is intended to bring forward the voices of teaching staff and students in order to complement the voice of institutional administration, and present a more complete and balanced picture. Theoretically, it is intended to put forward the perspective of teaching and learning to complement that of the organization and policy.

With the conceptual tool constructed above, the first research question would be investigated in the form of the three actors' perceptions of the three major elements in higher education, i.e. institution and higher education, teaching staff and teaching, students and learning. Regarding the second research question, these actors' concerns in their pursuit of quality can be directly raised as well as interpreted from their perception of quality of higher education.

4

Researching quality of higher education in China

This chapter aims at illuminating the overall rationality of the study and how it has been conducted. It first reports my methodological assumptions and choices. Then the overall design of this study is outlined, which is followed by the methods of my data generation and analysis, i.e. content analysis on 53 institutional self-evaluation reports, teaching staff interview and student focus group in three case institutions. Finally, I will also reflect on my methodology in terms of the reliability/objectivity, validity and generalizability of this study.

4.1 Methodological assumptions and choices

The aim of this study is to reconstruct a more inclusive and informed understanding on the quality issue of higher education. Thus, in the research I have tried to cover various experiences and multiple voices and to highlight the under-presented ones.

If we take the framework about research paradigms outlined by Guba and Lincoln (1994), the location of this study would be most close to constructivism, which holds (1) a relativist ontology: realities are alterable mental constructions, not something absolutely out there; (2) a transactional and subjectivist epistemology: the investigator and the object of investigation are interactively linked and the findings are literally created in the investigating process; and (3) a hermeneutical and dialectical methodology: individual constructions can be elicited and refined only through interaction between and among investigator and the respondents. However, I also assume that it is still possible and desirable to significantly document respondents' experiences.

Specifically speaking, in this study, I assume that the realities about quality of higher education are socially constructed (Berger and Luckmann, 1967) and context-dependent. They are apprehendable in the form of multiple, intangible and alterable mental constructions. And while the findings about quality of higher education are interactively created during the research process between me as the researcher and my informants of this study, they can still, to a large extent, document the experiences from the students' and teaching staff's perspectives.

Since this study aims at people's perceptions and concerns, which involves in understanding subjective experience and socially constructed meanings rather than testing hypothesis with relatively well-defined variables, the qualitative research methodology is more suitable than the quantitative one (Corbin & Strauss, 2008; Denzin & Lincoln, 2005). Qualitative research also allows me to focus on specific cases and use multiple sources of data such as interviews and documents. With case studies, I can 'investigate a contemporary phenomenon in depth and within its real-life context' (Yin, 2009, p. 18). Thus, I can engage more fully in interaction with my informants to elicit and distil a more informed consensus construction or understanding on quality of higher education.

4.2 Overall design of the study

As stated in the beginning of Chapter 2, China is selected as the context of this study mainly because only a limited literature on the quality issues in Chinese higher education can be found in the English language academic world, and my Chinese background offers me the opportunity to contribute on this. Besides, the trend of higher education policies showed a need to understand more about the quality issues in higher education.

Thus, this study started with a literature review and theoretical construction of the perspectives on quality in higher education. Based on the theoretical framework outlined in Chapter 3, this study limits and locates its empirical investigation to the organizational perspective and the educational perspective. For the policy perspective, I mainly resort to existing literature and sometimes may refer to related policy documents, however, the analysis and discussions will involve all the three perspectives. See Table 4.1 for the overall design of this study.

Content analysis of institutional self-evaluation reports was employed as the empirical basis for discussing the educational and organizational perspectives on quality of higher education at the institutional level. Altogether there were 53 reports being analyzed. Then individual interviews with teaching staff and focus groups with students were used to investigate their opinions and perceptions on quality of higher education, which forms the empirical basis for the educational perspective. 19 teaching staffs were interviewed individually and 45 students were interviewed in 12 focus groups. Those interviews and focus groups were conducted in three Chinese universities.

Table 4.1 Overall design of the study

Theoretical perspectives	Research method/resort
Policy	Literature and policy documents
Organization	Content analysis of institutional self-evaluation reports (53)
Education	Teaching staff interviews (19) Student focus groups (12, 45)

Based on the preliminary analysis of the research questions in Section 3.5 in the last chapter, the three actors' perceptions on quality of higher education are examined in terms of how they perceive three major elements in higher education – institution and the education it offers, teaching staff and their teaching, student and their learning. Table 4.2 has been developed to guide my empirical investigation. In the investigation I am trying to look for the respective quality-related attributes of the institution, teaching staff and student concerned by the institutions, teaching staff and students. Besides, the criteria they have used to judge these attributes have also been looked for. This is trying to formulate an interconnected, more sophisticated picture of the quality issue in higher education.

Table 4.2 Guide table for empirical investigation on quality of higher education

Quality of / Source/actor	Higher education institution	Teaching staff	Students
Institutional self-evaluation report			
Teaching staff			
Student			

Regarding the second research question about the three actors' concerns in their quality pursuit, it is sometimes a little bit sensitive to talk to a stranger directly about one's concerns or motivation, especially in the Chinese context. So they may not directly express their concerns. However, the concerns can still be extracted through interpreting the way they talk about their work or study. Concerns are also reflected in perceptions, as argued in last chapter. Therefore, in addition to direct questions about their concerns, I have looked into the actors' concerns in their pursuit of quality mainly by means of interpretation on their perceptions on quality in their respective contexts.

4.3 Content analysis of institutional self-evaluation reports

The official institutional documents formally represent the views of the institution as an organizational actor. Usually these documents are devoted to specific situations or issues. Proper analysis and interpretation on these documents in relation to their corresponding context is an important way to learn about the institution.

4.3.1 The role of the self-evaluation report

The self-evaluation report is produced by the institution during the ministerial evaluation on undergraduate teaching in China (cf. Chapter 2 for more information about this institutional evaluation). It is a formal way for institutions to communicate appropriate, relevant and up-to-date information about themselves to external world, specifically to the ministry and peer review group during the evaluation. The report not only serves as a basis as well as a major input for the review process by the evaluation committee, but also has presumably an important impact on the evaluation results. The institutions would reasonably perceive the results to have a significant influence on its reputation and on the perception of its efficiency and effectiveness in using public finances. So the institutions are bound to the extent possible to demonstrate their best work (according to their perception) in these reports. Therefore, a content analysis of these reports is a way to investigate the institution's perception of its quality and approaches in achieving it.

4.3.2 The method of content analysis

Content analysis is an empirically grounded method of examining texts, symbols and images in order to identify, classify and tabulate messages and meaning (Hartley and Morphew, 2008; Krippendorff, 2004). It involves not only describing what is said but also involves drawing inferences about the meanings in the messages (Holsti, 1969).

Inspired by Krippendorff (2004), Hartley and Morphew (2008), this study structured the following series of activities for content analysis:

- Sampling: Establishing clear criteria for selecting the institutional reports to be analyzed (elaborated on below).
- Unitizing: Identifying a set of discrete themes.
- Reduction: Systematically tabulating and summarizing data.
- Making inferences: Interpreting the patterns that emerge from the identified themes.

The themes to be addressed in the self-evaluation reports are actually prescribed in the ministerial Evaluation Plan. In this study, I have taken most of the themes directly from the reports, only changing some of their tiers and merging some closely related themes to make the analysis more operational. Then I put them into four categories that emerged from the analysis (see the findings in Chapter 5). Finally I have analyzed how the institutions addressed each theme and interpreted the patterns that emerged.

4.3.3 The sample reports

In the selection of the self-evaluation reports used in this study, I have tried to maximize the variety of institutions according to the frequently used institutional categorizations in China. My rationale is to try to identify common perceptions (if any) held by the diverse institutions in China. Then a necessary condition for inclusion was the availability of a report.

Ways of categorizing higher education institutions in China vary greatly and it is hard to find an approach that would be agreed by all universities. Categories frequently referred to in the Chinese context include reputation, mission, auspices, location and specialization. Due to the rapid development process, some Chinese institutions fall into several categories at the same time, while some are in the transformation from one to another, in particular, from specialized to comprehensive universities (Li et al., 2008). Therefore, classifications of Chinese universities are not always clear-cut.

The first round of five-year national evaluation on higher education institution in China was conducted in 2003–2008. Institutions were expected to post their self-evaluation report on the Higher Education Evaluation Center website (www.heec.edu.cn). However, many universities withdrew their online access after the initial evaluation. A general overview online search resulted in 53 reports as data resources for this study. This represents less than 10 per cent of all Chinese universities (Li et al., 2008,

p. 18) and there is hardly any way of ascertaining the exact representativeness of this sample.

However, the 53 reports include representatives of every category in Chinese university classifications that are most frequently used. It includes 24 top universities and 29 other institutions according to their admission in Project 985, which is an important symbol of prestige and government fund priority. Ten institutions are located in west region, 13 in central region and 30 in east region. According to their self-identification, based on the teaching and research focus, this study includes 21 research universities, 20 teaching research universities, four research teaching universities and six teaching universities and the remaining two did not clearly state their identification. There are 17 institutions under the provincial auspices and 36 institutions under the ministerial auspices. There are comprehensive universities and institutions that specialized in science and technology, agriculture and forestry, medicine and pharmacy, teacher education, art, political science and law, finance and economics, language and literature. Since more and more Chinese higher education institutions are in the transformation towards comprehensive universities (for example, Beijing Normal University has expanded from a teacher education focused institution to a comprehensive university), there is no definitive label in this sense.

4.4 Teaching staff interview and student focus group

My informants of staff interview and student focus group are from three Chinese universities, which are pseudo-named as University B, University C and University D. The pseudonyms B, C, D are used to the convenience that the acronyms of the three universities contain B, C, D respectively.

4.4.1 The case institutions for interviews and focus groups

In consistence to the research objective and the selection of self-evaluation reports, I have also selected these case institutions in the consideration of maximizing the variety in order to include various experiences and identify common perceptions of the teaching staff group and students group respectively. Or to use Yin's (2009, p. 54) terms, I have employed the logic of literal replication. Then there comes the practical consideration of accessibility, available time and energy, etc. Taking all these considerations, this study resulted in the following three universities for teaching staff interview and student focus group.

University B is one of the key universities in China, which is included and sponsored by both 'Project 211' and 'Project 985' (cf. Chapter 2 for more information about these two projects). It is a normal university located in Beijing and affiliated to the Ministry of Education (therefore, it's a *national* university). 'Normal' here refers to teacher education which is a highlighted section of programs provided by University B. But there are also a lot of non-teacher education programs covered in subject areas such as humanities, social science, natural science and engineering (therefore, it is a relatively *comprehensive* university).

University C is a university affiliated to Liaoning province of north China and thus a local university where a large proportion of the students are from Liaoning province (students from the local province accounts for nearly a half in the 2011 enrollment plan,[6] but it enrolls students from most provinces of the whole country). It covers subject areas mainly related to medical studies.

University D is a university affiliated to Guangdong province of south China and thus also a local university where most of the students are from Guangdong province (students from Guangdong accounts for more than 90% in the 2011 undergraduate enrollment plan, although it also enroll students from most provinces of the whole country). Most of the programs provided by University D are engineering programs.

Within each university, I have selected one of their typical programs for empirical investigation, i.e. the teacher education program educational technology at University B, the medical program clinical medicine at University C, and the engineering program automation at University D. And then I conducted my field work successively at University B, University C and University D during February and March, 2011.

4.4.2   Teaching staff interview

Through interview conversations, we get to know people and learn about their experiences, feelings, attitudes, and the world they live in (Kvale & Brinkmann, 2009, p. xvii). Interview conversations allow us to learn about people directly from conversations and understand ourselves as persons.

---

[6] The enrollment plan of regular higher education institutions can be acquired from the China Higher Education Student Information website: http://gaokao.chsi.com.cn/zsjh/.

4 Researching quality of higher education in China

With teaching staff interview, I am trying to understand their perceptions on quality of higher education in the context of their work and life at university. It was designed as a semi-structured conversation where knowledge is constructed in the interaction between the interviewer and the interviewee (Kvale & Brinkmann, 2009), i.e. me and the teaching staff.

Through the collaboration network between Aalborg University and the investigated universities in China, I got to know my contact person(s) at each university: two academics in University B, a vice dean at University C, and an administrative staff at University D. They had helped me in finding and contacting their teaching staff colleagues in the same college or school who had available time and wished to participate in my interview. I had asked my contact persons to look for staff who were teaching undergraduate courses at that time.

Altogether I have conducted 21 teaching staff interviews, with 19 of them audio recorded. One staff from University C and one from University D had refused to be recorded. I had taken written notes with these unrecorded interviews and taken some of the relevant points into the followed interviews. Table 4.3 shows an overview of the recorded interview participants (see Appendix 1 for more detailed information).

In the teaching staff interview, all the university B participants are teachers of the educational technology program. All the University C participants are teachers of the clinical medicine program. In University C, I also paid attention to covering teaching staff from both the clinical and basic sections of the clinical medicine program, where the teaching of the clinical section takes place in hospitals and the basic section in the classrooms and laboratories. All the University D participants are teachers of the automation program.

Table 4.3 Overview of teaching staff interview participants[7]

University	B	C	D	Total
Participants	7	6	6	19
Male	3	4	3	10
Female	4	2	3	9
Professor	3	0*	2	5
Associate professor	0	3	3	6
Assistant professor	4	1	1	6

---

[7] I failed to specify the academic position of two staffs from University C.

4.4 Teaching staff interview and student focus group

During my teaching staff interviews, the student affairs work of the mentor or form teacher was frequently mentioned. Several teaching staff straightly suggested me to interview the mentor or form teacher since they probably know more about student study and life at university. So during my field work I also interviewed one full-time student mentor at University C and University D respectively; but only the University C mentor interview was recorded (coded as CLA in my analysis), while the University D mentor had refused to be recorded. Besides, three of teaching staff participants themselves were mentors or form teachers at the time of my interview or had the experience as mentors or form teachers.

All the teaching staff interviews were conducted at the corresponding staff's own office. Before the start of each interview with the teaching staff, I introduced myself as a PhD student in the field of higher education who is interested in but knows little about their teaching, and who would like to learn from their experience. Then I told them that the interview will be used in my research report anonymously and asked them for permission to record the interview. And most of them said, 'no problem, just go ahead.'

Mostly, I started the interviews by requesting the participants to talk briefly about their work experience. Then the conversations turned to their teaching experience and the course(s) they taught at the time of the interview. Then I asked them about their perceptions, expectations and opinions on students and their study, higher education and university. The main interview questions are:

- Please briefly introduce your work experience and course(s) you teach.
- From your own teaching experience, what does a good teacher/teaching look like? Which aspects of your experience can be learned by other teachers, especially the new/young ones?
- What do you think about exams in relation to teaching and students?
- Among all the aspects of your work at the university, including teaching, research, etc., which aspects have you put most of your time and/or energy?
- Have you encountered any challenges in teaching? What kind of support do you need in order to improve your teaching?
- From your own teaching experience and interaction with the students, what do you think about the current students? What do you think they concern most? Which aspects do they put most energy? How is their attitude towards exams?

- From your own teaching experience, what do you think an ideal status is when a student reaches graduation? What does a good student look like?
- From your experience, which aspects do you think could show the quality of a university education? What does a good university look like?
- Do you have any other relevant ideas on what we have discussed?

Each interview lasted for about one hour. Although before the interviews I had checked the official websites and read the available materials about the university and the program, I tried to bracket my ideas about the topics and held myself an attitude of knowing-little or even 'naïve' to ask the basic and simple questions. I tried to relate to the participants as a curious student who was actively following up on their answers and seeking to clarify and extend the interview statements. After each interview I had made a brief reflection and noted down interesting themes that emerged and themes that needed to be verified in the following teaching staff interviews and/or student focus groups.

4.4.3 Student focus group

Focus group interview is a qualitative research technique used to obtain data about feelings and opinions of small groups of participants about a given problem, experience, service, etc. (Basch, 1987). There are several considerations in the use of focus group interview to investigate student perceptions and concerns on quality of higher education in China. First, it is, to a large extent, exploratory in the Chinese context; and focus group suits exploratory research very well (Vaughn, Schumm, & Sinagub, 1996, p. 6). Second, focus group offers students the opportunity to exchange opinions and perceptions based on similar study settings. During their discussion in groups they can inspire each other with their own words to develop more inclusive ideas about their study, teachers and universities. Third, some of the students may be too shy to speak openly to a strange researcher while more comfortable to discuss with peers. However, I am also aware that focus group interview might have some downsides such as when there are strong opinion leaders in the group the group dynamics may silence other individual voices of dissent (Kitzinger, 1995). Despite the imperfectness of focus group, it is still a useful method for my investigation with the students.

4.4 Teaching staff interview and student focus group

I got access to the student participants in University C and University D with the help of the same contact persons as my access to the teaching staff participants. In University B, one of my contact persons, an academic staff, had asked a student mentor to help me get access to the student participants. I had asked my contact persons or helper to look for one group of 3–5 students from each grade and it would be best if the students of the same group knew each other.

Then it resulted that at each university I got 15 students in four groups. Table 4.4 shows an overview of the participants in my focus group interviews (see Appendix 2 for more information about the student participants). There are four groups consisting of 3 students, six groups consist of 4 students, and one group consists of 5 students. Regarding gender composition, there are two all-male groups and one all-female group; the others are mixed groups. As can be seen from Table 4.4 there are only third-year participants at University C and only second- and third-year participants at University D; this is because only those students were available at the time of my field work or my contact persons could only get access to them.

Table 4.4 Overview of student focus group participants

University	B	C	D	Total
Participants	15	15	15	45
Male	8	4	11	23
Female	7	11	4	22
First-year	4	0	0	4
Second-year	4	0	7	11
Third-year	4	15	8	27
Fourth-year	3	0	0	3

All the university B participants are from the educational technology program. All the University C participants are from the clinical medicine program. All the University D participants are from the School of Automation. However, there are three engineering programs involved: three groups (D1, D2, D4) of the University D participants are from the automation program; and the other group from University D (D3) consists of two participants (D3a, D3b) from the 'Electronic Information Science and Technology' program and two (D3c, D3d) from the 'Electrical engineering and its automation' program.

All the participants of the same group were in similar study environment and knew each other before the interview. In most of the groups the participants were from the same class.

The focus group interviews were all conducted at a separate meeting room with no others present. Often when a group of students came to the room and met me there, they called me 'teacher' (which shows respect in the Chinese context), and I asked them to call me by my name and introduced myself as a student only several years senior than them. Then I told them my PhD thesis was about teaching and learning at university, and asked them for help to learn about their study at university. I promised to mention them in my thesis anonymously and asked them for permission to record the interview. And all of the participants agreed to be recorded.

Before turning on the audio recorder, I had also introduced how we would go on with the focus group interview: 'I will pose several broad topics or questions. If any of you come up with any ideas or comments with the topics or questions, just speak up. But you can say nothing if you do not come up with ideas. And you can also talk to each other about those topics.' When they had showed understanding about this, I would turn on the recorder and start the interview with the topic about their university and program selection, then their study at university, their learning experience, main concerns, activities, and then perception and opinions about exams, teachers, and university education. The main interview questions are:

- How did you choose your major/program and the university (after the higher education entrance examination)? Which factors did you take into consideration?
- When (in what kind of situation) do you think that you have learned the most? When do you feel fulfillment, satisfaction, achievement? What are the events or situations that have impacted you most in your growth and development?
- What are your main concerns? How do you spend your time and energy at university?
- What do you think a good student looks like?
- What do you think about exams?
- What do you think a good teacher/teaching looks like? Which *modus operandi* of your teachers do you think is good and could be learned by other teachers?
- What do you think a good university looks like? Which aspects do you think your university can improve to facilitate your learning better? If you were given the opportunity to re-select your study, which factors would be taken into consideration?

- Do you have any other ideas or comments on what we have discussed?

Each focus group interview lasted for 1–1.5 hours. I also tried to put aside my ideas and held myself an attitude of knowing-little to ask the basic and simple questions. I tried to relate to the student participants as a peer student who was curious about their study. After the interview I also shared with them my own study experience. After each interview I had also made a brief reflection and noted down interesting themes that emerged and themes that needed to be verified in the following student focus groups and/or teaching staff interviews.

4.4.4 Analysis of interviews and focus groups

As Kvale and Brinkmann (2009, p. 190) have suggested, the analysis of interview may, to varying degrees, be built into the interview situation itself. My writing a brief reflection after each interview or focus group formed the first step of my analysis. Then when I finished my field work, all the teaching staff interview and student focus groups were completely transcribed by myself. During the transcribing process I also noted down interesting themes that might be useful in my analysis, and what it could mean to my research questions. Since the interviews and focus groups were conducted in Chinese, I transcribed them also in Chinese. But my notes were a mixture with both Chinese and English, depending on which language came to my mind at the time of transcription. In the analysis, the themes were also noted down in Chinese, and when I wrote the report I used only English. I found that when the meaning was clear, the language difficulty was overcome by being sensitive to the nuances of different expressions and discussion with experienced researchers.

When the transcription was finished, I combined Kvale and Brinkmann's (2009, p. 201–218) suggestions on 'interview analysis focusing on meaning' and Corbin and Strauss's (2008, p. 45–64) approach of conceptualization by figuring out the properties and dimensions. First, I re-organized the transcripts into several word documents around the themes preset in my research design and interview questions, i.e. student and learning, teacher and teaching, university and higher education. Then I tried to look for the properties and attributes of the sub-themes (e.g. good students, positive learning experience or teaching experience) or to note down and classify the factors being taken into consideration in certain situations (e.g. university and program selection, time and energy allocation). These sub-themes are elaborations around the three main themes, which are partly preset from research design and which partly emerged during the interview. When this was finished, I then searched the potential inner connection within the teaching staff views as well as the student views; for example, I tried to look for the potential connection across the students' university and program selection, learning experience, main concerns, their perception of teachers and universities. Based on these inner connections, I re-constructed the students' and teaching staff's perceptions and concerns on quality of higher education. In these processes, I also integrated my reflection during the field work and my notes in transcription.

I have coded the teaching staff with a three-letter code; the first letter represents the university they are from, for example, CZC is a teaching staff from University C, DYL is from University D. And the students are coded by two letters with a number in the middle. The first letter also represents the university they are from and the first letter together with the number represents the group they belong to. For example, B3B is a student from Group B3, which is a group from University B; C1a is a student from Group C1, which is a group from University C. Besides, the case of last letter represents the gender of the student participants – upper case represent female, lower case male, for example, B3B, C2A are female, and C3a, D2a are male.

The difference between individual interview and focus group data has also been paid attention to. In the analysis of the focus group data, I paid attention to identifying areas of agreement and controversy, as a few researchers have suggested (Carey and Smith 1994; Sim, 1998). In the analysis, I attached individual ideas with an individual code, and ideas agreed by the whole group with group codes.

4.5 Methodological reflections

Here, I would like to reflect on my research process to offer my readers basic information for judging the reliability, validity and generalizability of this study.

I am a Chinese doing my PhD in Denmark and my study is in the Chinese context. This offers me the privilege to be inside and outside at the same time, to combine intimacy with the critical look of an outsider, involvement with detachment, as Bauman (2000) suggested in an essay on writing sociology. Compared to doing a PhD in China and study the

4.5 Methodological reflections 49

Chinese context, my experience of seeing something different in Denmark made me more sensitive to the issues I would otherwise take for granted. For example, before I came to Denmark I had thought that it would be mandatory for all universities to have a physical structure or a wall clearly separating the campus and outside world, but in Denmark there is no clear line dividing university area and non-university area. This produces internal validity in the form of sensitivity to the research issues and participants (Maaløe, 2009).

As I have already indicated in the introduction of the teaching staff interview and student focus group processes, I have kept myself at the ignorant position during my field work. This attitude has left my participants large space to conceive their ideas and come to terms with their experiences. In addition to this, I have tried to relate to both the teaching staff and student participants as a student. This gave them more comfort to express their perceptions and opinions. At the end of each interview I always asked an open question 'Do you have any other ideas or comments on what we have discussed?' I also made effort to clarify my understanding during the interview, to verify themes emerged in following interviews, to cover more voices and to build consensus understanding between me and the participants. These cautions are aimed at validation in the process and achieving objectivity and reliability in the forms of allowing the object to object and intersubjective knowledge (Kvale & Brinkmann, 2009, pp. 242–243).

As a person involved in human conversations, I also paid attention to the nuances of accent, gesture, face expressions to feel the authenticity of the interview and focus group. And when I was in doubt of this, I would push further with questions of 'Why?' or 'How?' to get down details to facilitate my interpretation. For example, since I got access to the teaching staff participants through their vice dean at University C, one of my participants there spoke to me just like spoke to the vice dean at the beginning of the interview, just saying everything was good. When I pushed the questions to concrete examples, he admitted that they had actually encountered significant challenges.

Being aware of the possible downside of the focus group method in silence individual voices of dissent (Kitzinger, 1995), I made a lot of effort to create an open group dynamic to encourage the student participants to speak up. In addition, I frequently posed the supplementary questions 'Any other ideas?' Mostly it resulted in a friendly atmosphere and often at the end of the interview, the students gave me a positive feedback to me that they had learned something new about their peers, learning opportunities and even themselves. And a University D student even said to me 'Why don't you use video-recording?'

With the available resources, time and energy, I have tried to maximize the variety of voices and experiences involved. The analyzed 53 self-evaluation reports are not perfectly representative for the higher education institutions in China. However, these reports still cover all the institution categories frequently used in China. The three case institutions for teaching staff interview and student focus group also cover three different types of institutions: a relatively comprehensive university with strong tradition in teacher education, a medical university and an engineering university; and they also cover the geographical north and south in China. The generalizing logic is similar to Flyvbjerg's (2001, p. 79) formulation of maximum variation cases: if such a variety of institutions, teaching staff or students have such and such perceptions and concerns, other institutions, teaching staff or students probably also have similar perceptions and concerns.

Closely related to generalizability, some researcher (Wellington, 2000, p. 138; Powney and Watts, 1987, p. 37) also mentioned the 'saturation point', referring to a situation where, after a certain number of interviews or case studies, ideas and issues begin to recur and reappear. In this study, many themes have reached the saturation point and a few may not. I have attached the participants' code to the ideas they held in the report of my findings, so that the readers can figure out how many times a certain idea has been repeated.

5
The organizational perspective[8]

The self-evaluation reports were produced by the institutions during the ministerial evaluation on undergraduate teaching (cf. Chapter 2 for more information about this evaluation). They were submitted to the Ministry of Education before the site visit of the peer review group during the evaluation. After the evaluation, they were made public on the website of the Higher Education Evaluation Center (www.heec.edu.cn). These reports have served as a formal way for institutions to communicate appropriate, relevant and up-to-date information about themselves to external world, specifically to the ministry. In this chapter, I present the findings from a content analysis on 53 self-evaluation reports.

5.1 The contents of the self-evaluation reports

The evaluation of higher education institutions in China was not carried out as a fitness-for-purpose process where the institutions set criteria for themselves. Instead, institutions followed preset standards and criteria formulated in the Evaluation Plan from the Ministry of Education. The institutional self-evaluation reports were prepared according to indicators from the ministerial Evaluation Plan. All self-evaluation reports consist of a brief introduction to the institution, which is followed by responses to the listed items requested by the Evaluation Plan. Some reports also include the activities related to the evaluation and a plan for improvement.

The indicators addressed in the Evaluation Plan, which formed the basis of themes covered by the self-evaluation reports, can be classified into four main categories – role of higher education, conditions of quality,

---

[8] The findings reported in this chapter have been previously published as Zou, et al. (2012). I acknowledge the Taylor & Francis Group (http://www.tandfonline.com) for granting me permission to reuse it here (permission reference: KB/CQHE/P8198).

approaches to achieve quality, and symbols of quality (see the Table 5.1 below).

Table 5.1 Categories and themes in the Evaluation Plan and self-evaluation reports

Categories	Themes in the Evaluation Plan and self-evaluation reports
Role of higher education	Institutional self-position and planning Thoughts/philosophy on education and the role of teaching
Conditions of quality	Teaching staff (number and structure, eligibility) Administrative staff (structure and quality) Teaching infrastructure (facilities such as buildings, laboratories, internships, libraries, intranet access) Finances
Approaches to achieving quality	Programme settings, curriculum, teaching through practice (workshop, laboratory) Quality control (regulations, standards and their monitoring and control) Teaching staff moral mentality (teaching staff's values, engagement, responsibilities, etc.) Learning atmosphere (students' conformation to institutional regulations, engagement and extra-curriculum activities)
Symbols of quality	Students' basic knowledge and basic skills Theses or designs for graduation/degree Students' ethics, cultural and psychological quality Students' physical health and sports activities Selectivity of enrolment Social reputation Employability

Through the items listed above the institutions are trying to create the following images of themselves in the self-evaluation report.

5.1.1 The role of higher education and higher education institution

All self-evaluation reports address the reporting institution's understanding of the role of higher education and universities in the sections on institutional self-position and planning. Usually the role is described and understood on the basis of research, teaching and societal service, with research holding the most prominent position. In the self-planning and positioning section, almost all institutions present themselves as being on the way to comprehensive (multi-discipline), research and high-level university. Five institutions among the six that identify themselves as teaching universities have clearly claimed their pursuit of research in the development goals, that is, to become teaching research university.

Besides, they claim to pursue a strategy of 'development through research'. Only two specialized institutions (one language institution and one art institution) did not mention their multi-disciplinary orientation.

All the 53 institutions also show that they emphasize teaching through the leaders' attention on teaching, sufficiency of teaching fund and resources and establishment of regulations guaranteeing well-functioning teaching.

5.1.2 Conditions of quality

Conditions of quality refer to the basic prerequisites for teaching and learning to occur; they primarily consist of staff (teaching and administrative), teaching infrastructure and finances. These conditions form the foundations of quality in higher education.

In the self-evaluation reports, the conditions are mainly described through quantitative indicators. The institutions calculate these according to the guidelines provided by the Evaluation Plan and then report the results.

All the 53 reports describe the teaching staff number, distributions of degrees held and distribution in technical positions and graduate institutions. The institutions show a preference for large proportions of young staff, staff with higher technical positions, staff with higher degrees and staff who have graduated from other institutions (especially those holding degrees from foreign universities). Furthermore, institutions also report on their policies for staff recruitment, training, assessment, promotion and incentives as well as list famous professors, professors engaged in key national programs, staff publications (even by the administrative staff) and awards given to staff.

As for teaching infrastructure and facilities, all the 53 institutions report the number of buildings, computers, multimedia classrooms and laboratories, bases for students' internship, books, intranet access, sports equipment and facilities. All of the above mentioned facilities are described as well as managed according various standard regulations. Also, awards acquired by staff and students based on the utilization of these facilities are described.

5.1.3 Approaches to achieving quality

Approaches to achieving quality refer to the processes and mechanisms employed to realize quality teaching and learning. This consists of evaluating academic management, quality control, teachers' morale, program and education plans and curricula.

Almost all the 53 self-evaluation reports employ examples to show the qualities listed below. In a few cases, where not all institutions are included, the figures in brackets indicate the actual number of institutions.

- Programs are updated according to social needs, meaning new programs (some of them interdisciplinary) are established and old programs are changed according to changing situations.
- Abundant numbers of courses have been selected as national or provincial Excellent Course Programs (52).
- There are detailed regulations for selecting textbooks and incentive for self-editing textbooks.
- Technology is integrated into teaching. Many courses use multimedia; some with self-developed courseware.
- There are incentives and assurance mechanisms for bilingual teaching by funding support and training opportunities.
- There are lists of awards for teaching acquired at the national and provincial level.
- Bases for students' internship and laboratories are well managed according to the corresponding regulations.
- Many awards are acquired and papers are published as a result based on the utilization of the internship bases and laboratories (45).
- Administrative staff are qualified in their degrees, technical position, training, management research and age.
- Quality is well maintained by following policies, regulations, procedures, standards and good-practice codes.
- Teaching staff are encouraged and obliged to work under good incentive plan and have acquired awards at national and provincial levels.
- Institutional regulations are strictly implemented. Awarding and scholarship are significantly used methods to encourage students to follow the regulations; and at the same time necessary punishment is used for students breaking rules.
- Students receive many awards for co-curriculum scientific and cultural activities. The institution supported these activities by institutional funding and preferential policies.

5.1.4 Symbols of quality

Symbols of quality refer to the signs that show quality teaching and education is achieved, including students' basic knowledge and basic skills, thesis or project design at graduation, health, citizenship, institutional social reputability, selectivity, awards, alumni and employment rates after graduation.

In the reports all the 53 institutions describe that:

- Students have received awards from a multitude of contests as symbols of basic knowledge and skills, innovative spirit and practical competence.
- Rigid and detailed regulations of thesis or project design for graduation have been well implemented, thus ensuring good quality.
- Students are healthy according to national standards. Many awards are acquired from sport competitions.
- Students obtaining high scores on the national higher education entrance examination have been enrolled from all over or the majority geographical areas of the country. The institutions thus present themselves as being highly selective.
- The employment rate for graduates is high. Most graduates have satisfactory jobs. This is often demonstrated through positive feedback from employers, though only 16 institutions formally presented employer survey results.
- There are successful alumni working in different areas in the society; and the institution has a good social reputation (36 institutions list examples of positive media reports).

Two particular and prominent elements stand out from the overall picture of the selected content in the self-evaluation reports. First, showing externally acquired awards is the primary way of demonstrating quality and is used by all 53 institutions. They try to show that staff, students, departments and the institution as a whole has continually won all kinds of external awards and honors. Second, each institution tries to show that there are comprehensive documents regulating all internal planning, codes of practice, frameworks, rules, procedures, regulations and policies of the institution's operation. The following is a closer look at these two elements and their implications.

5.2 Document-listing and awards as visible forms of quality

Institutions, by listing their comprehensive and detailed operation documents, try to present themselves as having a beautiful, ambitious future and often portray themselves as on their way to becoming a well-reputed, comprehensive research university by pursuing reasonably designed and rationally operated processes.

When it is difficult to demonstrate quality through documentation of results, the institution tends to show the documentation of a rigid process instead. For example, the thesis is one of the indicators under the category of teaching results in the Evaluation Plan. One institution spent more than two pages (out of the 85-page report) showing its detailed stipulations on the organization of thesis related work such as theme-selection, supervision, assessment and defense. Details were also given on the job divisions between leaders, teaching staff and administrative staff. The requirements also specify writing procedures, the text format, specific content and design of the thesis cover. Supervisors are required to hold a higher technical position than assistant professor and each supervisor can supervise no more than five theses.

The assumption may be that rigid process regulations guarantee quality outcome. In practice, however, this may not be true for the actual process. As natural system analysts in organizational studies have emphasized, there is more to an organizational structure than prescribed rules, job descriptions and the associated regularities in the behavior from participants. Individuals are never merely 'hired hands' but bring along their heads and hearts: they enter the organization with individually shaped ideas, expectations and agendas and bring with them distinctive values, interests, sentiments and abilities (Scott and Davis, 2007, p. 63).

However, actual processes are not as easy to document as listing documents. Planning, policies, regulations and procedures are means to achieving the institution's goal. However, there is the risk that goals and means may be disconnected when members of the organization are rewarded for their ability to stick to procedures rather than their contribution to achieving more general goals. With their emphasis on adherence to rules, organizations are vulnerable to goal displacement (Merton, 1957, pp. 199–202), a process in which goals are forgotten as members of an organization come to see following the rules as an end in itself.

Besides document listing, all the institutions list their externally acquired awards to show that they have been externally recognized. This is

5.2 Document-listing and awards as visible forms of quality

a way of demonstrating their quality; but seen from the perspective of teaching and learning, it is a very limited view.

Each of the external awards, usually given by different agencies, only focuses on one or several specific aspects of the institutions, staff or students; it can be hardly expected that these awards would systematically and consistently align with each other or with the work of higher education institutions. By referring to awards as a sign of quality, it is not clear what a higher education institution should be and why it exists. The essence of such awards is their relative position among the institutions according to the focus and criteria set by a certain award, containing no substantial reference points to address the intrinsic values of education. Awards serve as only a demonstration, a symbol or a by-product, of education, not education itself. Emphasis on awards give limited implication of a systematic approach to institutionalized work and provides little information on how to improve. For example, most reports give more weight to the description of what prestige awards the students have achieved than how they achieved the success. The latter can give further reflection on the efficiency of teaching and learning methods. Furthermore, the contest assessment criteria are not necessarily consistent with the curricular requirements of most of the students. The related educational program has a much wider scope than the assessment criteria of the contests.

In addition, awards involve only a very limited proportion of students and staff and can hardly reflect the overall or average level of quality. Awards target the best students, thus activating only those few people who have the potential or prospect of competing.

As regards the teaching and learning perspective on quality in higher education, as previously discussed in this paper, this study observes that teaching and learning has not been the main focus in the self-evaluation reports despite the goals of this evaluation system, as stated in the Evaluation Plan. Little, if anything, is said directly about the elements of the educational process or about quality models, quality enhancement and quality feasibility in teaching. Instead, it seems that the construction of the reports followed the logic of visibility and social recognition showing the most visible and prestigious parts of the institutions. The internal document listing and reliance on awards to show quality vividly illustrates this logic. Actual teacher–student interaction remains invisible and the intrinsic logic and value of education is seldom addressed.

However, looking beyond the higher education institutions' engagement in education and regard them as organizations located in a certain social environment, the content of the self-evaluation reports becomes more understandable.

5.3 Higher education organizations and their environments

From the organizational perspective, as articulated in Chapter 3, all organizations must pursue or support 'maintenance' goals in addition to their output goals (Gross, 1968; Perrow, 1970, p. 135). Institutions of higher education, like other organizations in society, are struggling for survival by raising resources from the environment. An effective organization is one that responds to the demands of its environment according to its dependence on the various components of the environment (Pfeffer and Salancik, 2003, p. 84). The government, represented by the Ministry of Education, is the main resource provider to the higher education institutions. In other words, the institutions depend on the ministry. Thus, it is important for institutions to be recognized as effective and efficient by the ministry in order to be able to apply for more resources.

However, it is not easy for institutions of higher education to evidence their effectiveness and efficiency. Ideas about organizational efficiency and effectiveness, in the public policy discourse as well as in mainstream organization theory, are generally modeled on companies in the private sector. Such business organizations are confronted with two types of demand; they have to deliver products or services that are sufficiently useful and make them available so that consumers will buy them and they have to provide sufficient profit to be a trustworthy object for investors. The second characteristic of businesses is most pertinent here. The trustworthiness of a business organization is measured in money and, hence, profitability becomes a main criterion for internal decision-making.

Higher education institutions and companies of comparable size are both complex organizations and may share many characteristics. Comparisons of private and public management often tend to assume that management of private companies is much more rational and efficient than management in public institutions; but this assumption is far too simple. Many companies as well as universities may be described as organized anarchies for the basic reason that they depend on groups of people with specialized skills and some measure of autonomy to deliver their services. The main difference is that companies can (and have to) use the profitability criterion in internal decision-making, for instance in closing down or expanding activities, while higher education institutions are confronted with other and more complex demands (Clark, 1983; Birnbaum, 1988; Cohen and March, 1986) and have to use more complex criteria. As public organizations, the mission of higher education institutions is decided by state policy; mostly it is to provide quality education to a certain group of people and provide quality research. It is not impossible to specify these goals. For instance, students' completion of studies on time may be specified as a main criterion for education; and publication of articles in certain types of scientific journals may be specified as a main criterion for research. However, there are other criteria as well, which are often not well defined. As institutions with a public and politically constructed mission, higher education institutions cannot reduce the quality of their services to the criterion of profitability. Thus, the difficulty for higher education institutions is two-fold: the demands on them as public institutions are often ill-defined and they cannot use the criterion of profitability in internal decision-making in the same way business organizations do.

These problems make it very difficult for higher education institutions to demonstrate their efficiency and effectiveness directly causing them to turn to socially visible and legitimated elements: awards and internal document-listing.

Award giving, showing someone's relative position in a reference group, is thus a strategy employed by higher education institutions to demonstrate their quality in a situation where standards of desirability (goals) are ambiguous and cause and effect knowledge (technology) is believed incomplete (Thompson, 2003, pp. 84–87). The institutions' broad goals such as teaching, research and social service can hardly be used to guide their daily operations or show their effectiveness, efficiency or quality. Furthermore, it is not clear which approaches or tools can be used to achieve the position of a good or even world-class university. Instead of facing these difficulties and risking being the focus of controversy, higher education institutions have employed the socially legitimized/recognized awards system to show their quality 'safely' compared to other institutions (their reference group). In the discrete aspects, the institutions place significant emphasis on corresponding awards to create an image of being good to the outside world. From this organizational perspective, the elements focused on in the self-evaluation reports appear to be logical means for gaining visibility and social recognition.

5.4 Summary of quality from the self-evaluation report point of view

Seen from the self-evaluation reports, the institutions' concern on quality regarding student, teaching staff and institution as a whole can be summarized as following:

- **Students' awards, employment rate**

    The concern on students' awards appeared in three out of the four themes in the institutional self-evaluation reports. Under the theme *'conditions to quality'*, there were student awards acquired based on the utilization of infrastructure and facilities. In *'approaches to achieving quality'*, there were student awards based on the internship centers and laboratories, and on the institution initiated or supported co-curriculum scientific and cultural activities. Besides, there were institution-arranged awards and scholarship to encourage the students to study and follow the regulations. As *'symbols of quality'*, student awards (in a variety of academic contests and sports competitions) were more directly presented to demonstrate students' basic knowledge and skills, innovation and practical competence, and health condition. The employment concern is mainly shown in *'symbols of quality'* by reporting the employment rate and listing successful alumni working in different areas in society.

- **Teaching staff's awards, publications, research projects, reputation**

    Teaching staff's awards highlighted both *'conditions of quality'* and *'approaches to achieving quality'*. In *'conditions to quality'*, staff awards were directly presented in staff profile and also mentioned in facility utilization. In *'approaches to achieve quality'*, awards for teaching and teaching staff at both national and provincial level were directly listed, and also mentioned in addressing the laboratory utilization. Publications and research projects, especially national key projects with large amount of funds, also highlighted those two themes in staff profile and laboratory utilization. The reputation concern is mainly shown in the self-evaluation reports' highlighting famous professors, faculty members holding key positions or engaged in national key programs.

5.4 Summary of quality from the self-evaluation report point of view

- **Institutional comprehensive operational documents, awards, reputation, quantities of infrastructure and facilities**

The concern on an institutional image of comprehensive and well implemented operational documents can be seen from all the four themes in the self-evaluation report. These documents include institutional self-positioning and planning addressing *'the role of higher education'* and the specific institution, and all kinds of codes of practice, frameworks, rules, procedures, incentive plans, regulations and policies in the other three themes. The institutional awards concern is directly shown in listing awards acquired by the institution or its departments in *'approaches to achieve quality'*, and also in their listing of courses admitted in National Excellent Course Program, which is also award-equivalent in the sense that it is conferred a state of high honor with financial support. The concern on institutional reputation is clear in the institutional planning self-positioning as or to be comprehensive research university and the highlighting of national key programs and laboratories held. Quantities of infrastructure and facilities were concerned mainly in reporting their *'conditions of quality'*.

This study undertook a content analysis of 53 self-evaluation reports from Chinese higher education institutions in order to examine how they understand and demonstrate their own quality. The analysis and discussion show that higher education institutions, in general and worldwide, have both educational and organizational dimensions to their quality and find there to be some tension in the pursuit of these two kinds of quality, even though they are somewhat intertwined. In the pursuit of educational quality, higher education institutions are devoted to their educational arrangements and their impact on the students' mind (Barnett, 1992, pp. 6–7). Thus, the focus is teaching and learning activities. However, in the pursuit of organizational quality, higher education institutions are devoted to being socially visible and recognized as 'good' in order to gain the resources and legitimacy for their survival and prosperity. Thus, the self-evaluation reports are mainly a means to the end of organizational quality. This study shows the potentially limited and even distorting effect of this type of evaluation arrangement on the educational quality of higher education institutions, that is, their teaching and learning quality.

6

The teaching staff perspective

In the previous chapter, the educational perspective was only dealt with theoretically; in this chapter and the following one, I will report my empirical findings from the educational perspective, i.e. the points of view of the teaching staff and students.

The teaching staff in higher education are the frontline workers who directly interact with the students and facilitate their development. As a group, they also face the institutional administration where they get resource support for their work and incentive for changes. Their views on quality of higher education form an essential basis for understanding quality educationally. This chapter first reports the findings from 19 teaching staff interviews around three broad topics, i.e. teaching and teachers, students and learning, university and higher education. Then a preliminary analysis is followed to highlight the emerged themes concerned by the teaching staff, which I regard as quality of higher education from the teaching staff perspective.

6.1 Teachers and teaching

Around the 'teaching and teachers' theme, I have investigated the teaching staff's views on their desired outcomes of teaching and education, opinions about good teaching, challenges and endeavors in their work and their expected support, and their opinions about examinations.

6.1.1 Desired outcomes for teaching and university education

When teachers talked about the objectives of their current courses and their opinions about the ideal status of graduates, they mentioned a variety of aspects of the students that can and should be developed as a result of their teaching in particular, and university (undergraduate) education in general. These aspects include the mastering of basic knowledge and basic skills, ability to transfer knowledge and learn by oneself, and other personal and professional developments.

6.1.1.1 Basic knowledge and basic skills

Most of the teaching staff I interviewed put emphasis on the students' mastering of basic knowledge and basic skills of their corresponding subject, discipline or profession (BLM, BCG, BMN, BCL, CFC, CLA, CLN, CWB, CZH, CZC, DCW, DGJ, DXY, DZQ). Basic knowledge consists of the basic theories, concepts and methods (procedures) of the corresponding subject or discipline and basic skills consist of both skills in general, such as logical thinking, making inference, calculation, communication (written and oral) and scheduling, and skills specific to their subject or profession, e.g. modeling in discrete mathematics, clinical operational skills for the medical students, and practical engineering skills for the engineering students. But the roles of this knowledge and these skills, and the approaches to achieve them, may be quite different in different disciplines or professions.

To the teachers, mastering the basic knowledge, methods and skills of the corresponding discipline or subject is one important aspect in undergraduate learning. This is the foundation for discussion, further learning or application.

The emphasis on basic knowledge is most typical from the University C teachers. All the medical teachers interviewed emphasized the importance of basic knowledge. Memorizing basic knowledge seems to be one of the basic approaches of medical study. Several teachers stated very explicitly that memorizing basic knowledge is the first step for medical study on which discussion and further study or application are based (CWB, CFC, CLA). They also admit that there are too many courses and too much knowledge content for the medical students to memorize. So, as graduates of the clinical medicine program, the students should be able to remember or know the index or structure of knowledge so that they know where to look for relevant knowledge when they encounter a specific disease or patient. This is well illustrated by CWB and CFC in the following quotation.

To study medicine, memorizing is the most basic way of learning. I feel that at the beginning it's just rote learning (memorizing without any understanding). And then when you come to the clinical course and see a patient, you may feel something related to what you have memorized. Medical students are very good at memorizing: a thick book can be memorized in several days. They are forced to be like this. Memorizing is the basis. The basis of studying medicine is to remember. If you haven't remembered the basis, how can you use it? CWB

And another teacher from University C, CFC, has a similar opinion:

Biochemistry looks more of science (which is more based on understanding and logical thinking and speculation), but actually much of the contents is beyond discussion, i.e. the basic theories. For example, the sugar pathway is the sugar pathway, you don't need to discuss it; what you should do is just memorize it. Many courses of medicine, including the clinical courses, teach you the basic knowledge. You can discuss, but the discussion should be based on these bases, and what you should do at this stage (of study) is memorize. CFC

Then she went on and explained the function of discussion.

The discussion I am talking about probably doesn't happen in the student period but in the period when the students have mastered the basic knowledge and [then they can] discuss the treatment of some diseases. ... The knowledge level of the students makes them incapable of discussion. ... In class I may give them some exercises to discuss. But this kind of discussion is not discussion about basic theories; it's an exercise of using the basic theories to explain certain disease syndromes from the biochemical perspective. CFC

Although the University C teachers also emphasized the importance of students' mastering of clinical manual skills, the emphasis on basic skills is best illustrated by the University D teachers. They put great focus on the students' practical engineering skills (such as making silicon chips, designing and maintaining a production line, using professional software, etc.) which enable them to get real work done, to perform well in engineering projects (DYL, DCW). Besides their professional knowledge and manual skills, performing well also requires the students to be competent in communication such as writing reports, giving lectures, designing posters, and working in collaboration with others (or working in teams) (DGJ, DZQ). For the students to achieve these skills, it is best to give them opportunities to participate in projects, be it the teacher's projects, or certain kinds of contest (e.g. the National Electronic Design Contest), or internships in companies.

DGJ from University D also differentiated between emphases on theoretical knowledge and on practical skills for students who have different career orientations. For those students who go directly to work after graduation, the emphasis is more on the practical skills which are needed in the work; and for those who pursue further education with higher degrees, the emphasis is more on the theoretical knowledge, which will be tested in the entrance examination for graduate education.

Besides the subject-specific skills, the mentor teacher from University C, CLA, also described the ability to schedule, to arrange different issues and make plans as an important aspect of good graduates. He said that good students can schedule very well for their course study, examinations, thesis, job-hunting, etc. These students have good plans and know how to prioritize all kinds of issues. In addition to knowledge and skills, one teacher from University D, DZQ, proposed that interpersonal competence and the ability to appreciate artistic works, music, literature, etc. should also be developed during the students' university period. But he regarded this development as more dependent on the students' personal interests than the responsibility of teachers or universities.

6.1.1.2 Ability to transfer and learn

According to the University B teacher participants, the information technology (IT) courses are more oriented to the students' ability to transfer (e.g. BMN, BZG). The development of and change in technologies is very fast. Students more and more frequently use all kinds of technology. The learning of a specific technology is aimed not only at the mastering of that technology, but also at providing the students with experience of learning something new and paving the way for learning technologies. When the students encounter something new in the future, they quickly get to know how to proceed by connecting the new situation to what they have learned. For example, the Java teacher from University B formulated his course objective as follows:

Take the Java course for example, I think the most important thing is to [let the students] learn the way of thinking in programming. Because the programming languages are changing all the time, they are different every year; first it's Pascal, and then C, C++, and soon it's Java, and now a lot of languages derive from Java. So I think, for the students, the most important thing is to learn the way of thinking in programming, and then he/she can learn a new language very fast. BZG

The ability to transfer can be generalized and raised to learning ability in general. As CWB, a teacher from University C, said,

The graduates should have learning ability; and this (criterion) can also apply to (the students of) other professions or disciplines. [Learning ability is the ability that] you can [learn to] do something you've never done before to the extent that, although not as well as experts, it definitely exceeds people's expectation. CWB

DXY, a teacher from University D, made a comparison between the graduates of occupational or technical education and those of university undergraduate education. In her opinion, it is important for students to be able to solve new problems independently. She said that:

The occupational and technical education [only] enables the students to do the operations needed in the frontline; and the university undergraduate education not only enables the students with these operational skills, but also enables them to solve new problems and solve problems independently with the theories they have been equipped with. The theories may be useless in usual situations, but when some problem occurs, they probably provide the students with opportunities (to demonstrate their ability) unreachable for others. DXY

The ability to transfer and learn enables the students not only to apply the learned knowledge and adjust to the new specific situations, but also to learn and acquire new knowledge based on the learning experience they already have.

6.1.1.3 Other desired outcomes
Besides the above two widely desired outcomes from their teaching, the teachers also indicated a number of other characteristics that they would like their graduates to possess. They wanted the graduates to be clear about and have confidence in their future career development, to internalize the ethics of their profession and to gain external recognition in education or in the labor market.

Several teachers (CFC, CZH, DXY, DCW) also mentioned that good graduates should also be clear and confident about their career development based on their training in the corresponding discipline or profession. Based on their basic acquaintance of the profession or discipline, the students should first know their own position within it and be able to choose a direction for future development. Then they should also know what they have learned and are able to do, and be confident that they can do it well. Finally, they should be clear and enthusiastic about what they are going to do. In line with this, it is to the teachers' delight to have their students accepted by more prestigious universities for further study or accepted by better employers as employees, which they take as external recognition of their students (CLN, CLA, DXY, DYL, DZQ).

One teacher from University C, CLN, emphasized the importance of professional ethics in teaching medical students. He said that:

Doctors are facing real human beings where it is easy to make irrevocable mistakes, so the requirement for prudence is higher than other professions. Thus in medical teaching the teachers should cultivate the students to be prudent, realistic, serious, and responsible for patients and for themselves. CLN

However, some medical teachers (CLN, CWB) emphasized that the training of a doctor cannot be completed within university education, or at least not within the five years of undergraduate education. As CLN said,

There still needs some time for medical program graduates to become mature doctors. The graduates from the five-year clinical program are like some preliminary products. They have to relearn in detail the knowledge of the department in which they start their work. And they have to get a mature doctor to guide them in differentiating treatment in different situations. Experience is a vital part for doctors. CLN

6.1.1.4 Summary of the teaching staff's desired outcomes

To sum up, the ideal teaching and education results from the teacher's perspective of having equipped the students with knowledge and skills in both their corresponding subject and beyond, with the ability to learn, and made the students interested in the discipline and clear about their career in the profession. As an external recognition, the teaching staff also hope their students get better employment or are admitted by prestigious institutions for further study.

6.1.2 Good teaching from the teacher's perspective

When they talked about their opinions of good teaching, good teachers and the aspects of their own practice that they regard as good and that may be shared with other teachers, especially the young and new ones (lessons they learned from their experience), the teachers expressed their currently desired ways of teaching (quality teaching) from their perspectives. When these topics were raised, they mainly talked about teachers' interaction with students, dealing with teaching contents, popular teachers' personal attributes, etc.

6.1.2.1 Teaching experience in relation to the students
When they talked about their teaching experience in relation to the students, the teaching staff mostly mentioned the interaction with students both in and outside class, sharing with students some experiences and examples, and offering students the opportunities to learn by doing.

- *More interaction in class*

    Interaction in class here is set in contrast with a teacher talking all the time during class time; and it mainly refers to the mutual raising and answering of questions (or discussion) between students and the teacher (BCL, CZH, DCW), and sometimes also refers to the teacher's paying attention to students' response or feedback from their facial expression or eye contact when they are teaching, and adjusting teaching accordingly (CWB, CZH), or giving students opportunities to present their work in class and comment (BMN).

    Interaction in terms of the mutual raising and answering of questions in class can be used as a tactic to mobilize students, to force students to think, and to promote students' understanding (BCL). It also helps teachers to learn more about students' understanding of knowledge (CZC). The responses of students (facial expression, eye contact, etc.) in class show whether the students understand or not and whether they are interested in what the teacher is talking about. This gives clues to teachers for adjusting their teaching. The young new teachers, in comparison to the experienced ones, usually stick too much to previously prepared contents and are not so able to use these clues to adjust teaching flexibly (BCL, DCW). The student response is even used by some teachers (CWB, CZH) as an important aspect for assessing teaching and important sources of satisfaction and achievement. As they said,

6 *The teaching staff perspective*

If I could decide on how to assess teaching, I wouldn't use the exam scores; I'd rather use the extent to which the students' interests are attracted. CWB

If there is no response from students when I finish a class, I feel that class is just a failure. CZH

And CZH added that

It's very easy to just talk and just focus on the contents in class, but then you wouldn't get responses from the students. It's more tiring to try to mobilize students, to interact with them. But when you are trying to interact with the students, it's a symbol to them that you care about them, and they will give a response even if they are uninterested in the contents. So if you talk about how good a class is, to me it's the response of the students. CZH

And for new inexperienced teachers, who are not able to master the class dynamics like the experienced ones, it is a strategy to avoid boring classes (BCL). In addition, questions can also be used at the end of a class for the students to summarize or reflect on what they have learned in the class, or to provoke interest in what they are going to study in the next class (CFC).

- *Sharing experiences and examples in teaching*

Sharing experiences and examples is also an important teaching strategy for the teachers (BCL, CMM, DGJ, DZQ). The experiences and examples shared can be from the teaching staff themselves, from the students or from others, only if they interest the students.

BCL, a young female teacher from University B, portrayed one aspect of popular teachers as communicating with students about what they (the students) are concerned about with vivid cases; she said that:

Popular teachers may integrate in class some issues that students care about or are confused by, and talk about some related cases. These issues or cases can be situations encountered by the teachers' former students, or just something that recently happened. These may [and may not] have something to do with the teaching contents. The point is that the teachers communicate the information; and this makes them popular with the students. The students have probably learned from this type of class not only the contents listed in the syllabus but also something else such as ways of learning, ways of thinking, ways of communication, or how they can induct in the university, how they can apply what they are learning. BCL

She reflected that this way of teaching vivifies the contents taught with flesh and blood by providing students with specific examples which they are familiar with, putting the contents into the context of the students' lives and thus bringing them alive.

Besides communicating with students other students' cases, some teachers also regarded sharing their own experience as a good way of teaching (BCL, CMM, DGJ, DZQ). Some teachers shared with students their own experience of being a student. As CMM said to the students,

When I was a student, I found this or that really difficult to learn or memorize. And I overcome the difficulty in such and such ways, for example, by using doggerel like this ... CMM

This makes both the teacher and the contents more approachable.

Engineering teachers may share with students their own experience of conducting real engineering projects (DGJ, DZQ). They provide the students with the project context, and problems or objectives, put the students in the position of project manager, and then ask them to plan for the project, and discuss factors that should be taken into consideration. Finally, they show how the project has actually gone through and the effect of the project. Sometimes they also share with students the problems that they encountered in some projects and ask students to think about what can be done in those situations, and then show the students what they have thought and actually dealt with the problems. This experience-sharing approach vivifies the abstract theories in the textbooks and makes them interesting to students. In addition, it also shows students that what they are learning is useful.

- *Offering students opportunities to learn by doing*

Many teachers mentioned the importance of providing students with opportunities to solve problems, conduct tasks and projects, so that they can learn from their own experience (BLM, BZX, CZH, DCW, DGJ, DXY, DYL, DZM, DZQ). In these experiences, the students can put what they have learned into practice or application and learn some first-hand knowledge, which also motivates the students. This also provides opportunities for peer learning. BZX, from University B, shared her ideas about the importance of student experience in teaching. She said that:

I prefer to organize my teaching in the form of activities. The knowledge cannot be directly [transferred] from the textbooks. The students have to experience, and then combine experience and the propositional knowledge. Only this combination leads to real learning. BZX

CZH experienced the motivating effect of the problem-based learning (PBL) approach and said that:

When they had got used to the PBL way of teaching, the students showed great curiosity in knowledge. Some of them became extremely interested in the knowledge when they could actually see the connection between the knowledge in cell biology and clinical cases. Then they were more active in asking questions, thinking about the disease cases. One of them even said to me that he felt more like a doctor, no longer studying just for examinations. CZH

When they are motivated by the opportunities to do something in the learning process, students tend to be active in learning and will explore different sources of knowledge themselves.

- *Informal contact with students*

Some young teachers also talked about the experience of trying to be close to students (BMN, BCL, CWB, DGJ). Being close to students, or even one of the student community, makes the teacher popular with the students, and thus is good for teaching to go through. Sometimes the students like a course just because they like the teacher (BCL). If they like the teacher, they will care about the course(s) he/she is teaching (CWB). As BMN said,

From my experience, I think teachers should be members of the (student) community, and interact with them in a more relaxed way, even make jokes with them, and not make students feel that I am a teacher and you are students, and we are different in age. So the students will be fonder of us, close to us; and this makes teaching easier to go through. BMN

Good teachers, in some teachers' opinion, are the ones that love their students but are strict with and take responsibility for them. They are very kind to students as they are with their kids; for example, they would kindly remind students of some practical issues (e.g. not forgetting their textbook) (CFC). But it was also emphasized that this love should not spoil the students; and this love wouldn't result in lower requirements of academic rigidity or fewer assignments (BLM). Rather the teacher should still stick to the criteria of what you are supposed to know or be able to do after the course, or even raise the requirements or standards to make sure the students stand at a more advantageous position in the job market (DXY). In the teachers' perception, the students know and understand that the teachers are there for their (the students') own good.

Some teachers (DXY, DZM) have kept informal personal contacts with the students after their graduation. They take this as a way to gain feedback from alumni about the (social) relevance of their teaching in the students' work, and the change of societal requirements for university teaching. Then they adjust some aspects of their teaching accordingly.

One of the University B teacher, BMN, also shared her positive experience of using information technology to combine academic and non-academic issues. She created an online forum where she and her students could have discussions. She divided the forum into two main sections, one is course-related where there are several course themes they can discuss, and the other one is a social one where the students can share everything they want, for example jokes, or their feelings about the university, etc.

6.1.2.2 Teaching experience in relation to teaching contents

In addition to their experience of interacting with the students, the teaching staff also reported their experience of dealing with contents in teaching. Their highlighted experiences included making the contents interesting to the students and giving overviews.

- *Make the contents interesting and the students interested*

When the teachers talked about dealing with teaching contents, they mostly mentioned relating the contents to students in a variety of ways, in order to let students make sense of and be interested in the contents.

Many teachers (CFC, CZH, CWB, DGJ, DYL, DZQ) emphasized the importance of activating students' interest in the contents. For example, DZQ from University D said that:

Interest is of most importance in the learning process; if a student is totally uninterested in the contents, it is basically of no use even if you force him/her to learn. DZQ

To make students interested, one approach is to relate the contents to something the students are familiar with or visualize abstract contents to make the contents interesting. In the University C medical program, the teachers talked about relating abstract theories to disease or patient cases, and clinical operations, sometimes also to daily life (CFC, CZC).

This course, pathophysiology, is about the mechanisms of diseases; and it is kind of boring if you only focus on the mechanisms. The students would be more interested if you could relate the mechanisms to some disease cases. For example, the mechanism of heart failure can be related to a patient case with hypertension. You can show the X-rays and the students will see the swelling of the heart. Then you activate their curiosity and start to explain the mechanism. CZC

At University D, the teachers tried to relate the contents to real application, engineering projects that had been actually implemented, so the theories are brought to life in real contexts (DGJ, DYL, DZQ). Some University D teachers have also used computer simulation to visualize some models or processes (DGJ). University B teachers mentioned quickly coming up with examples as a characteristic of experienced teachers (BLM, CFC).

A second approach to making the contents interesting to the students is to let the students see the usefulness of the contents and the connection with their future (work, further study, etc.) (BCL, BCG, CFC, CZH, DGJ, DXY, DZM), for example, showing the usefulness of the subject contents in explaining the mechanisms of diseases, the clinical work, the real application in engineering projects, and in the entrance examination for graduate education.

The third approach to making the students interested is relating the contents to the research frontier, a recent application, or some hot topics (e.g. Nobel Prize) to interest the students and to show them possible options for future development (CFC, DXY).

Sharing with students other students' or the teacher's own experience of learning, solving problems or conducting projects, as mentioned in the above section, is another way for teachers to interest students. Some teachers (CZC, CMM, BCL) also mentioned making beautiful and illustrative courseware (e.g. PowerPoint) and being humorous as ways to interest the students.

- *Give overviews*

Another aspect of dealing with teaching contents mentioned by the teachers is giving overviews of the contents and the characteristics of the course or program, and how they are organized (BCL, DYL, DZM, DZQ). First, the students should be introduced to having an overview of the program or discipline in the first semester, which consist of information such as what this program or discipline is about, what options could be for the students' future development or work by studying this program or discipline, what elements or types of course the curriculum consists of, what the basic contents and objectives of each course are, what the connections between different courses are, etc. The head of the automation program at University D, DCW, gave an example at the curriculum level. She said that:

We have actually had a course called 'Program and Discipline Introduction' since 2000; and I am in charge of this course as the program head. We have invited some famous professors of the program to give introduction lectures on different aspects of the program together with me. The course is aimed at familiarizing the students to the program and discipline and cultivating their interests towards the discipline; and the course introduces them with what the program and discipline are about, which areas they are committed to, and what the recent technology developments in these areas are. DCW

For each course, the teacher should also give students an overview of the course at the beginning (BCL, DYL, DZM, DZQ). The teacher should make clear the role of the course in the program curriculum, the overall plan and arrangement of the course, the connections with other courses, the contents, objectives and characteristics of the course, and the corresponding study methods, etc. And in the teaching process, the teacher should pay attention to helping students to build their knowledge structure (especially connections with future courses and experiments).

6.1.2.3 Teachers' personal attributes

It seems to be difficult to talk about good teaching without talking about good teachers. One of the teachers, BLM, who is teaching a course titled 'Teaching Design', stated explicitly that 'good teaching firstly needs a good teacher'. Now let's take a look at the personal attributes of good teachers from the teacher's perspective.

- *Commitment to education*

Good teachers should commit themselves to the education enterprise, love and put their hearts into teaching, and care about and be responsible for the students and teaching (BLM, CFC, CWB). They should spend enough time and energy in teaching (BCL, BZX, CLN, CWB, CZH, DGJ) and should seek improvement in teaching even if it is not so encouraged as research by the institution. As DGJ said,

We teachers should improve our teaching even though the current university policy does not put so much weight on teaching in the teacher evaluation and promotion system. DGJ

- *Mastering the teaching contents*

In the teachers' opinions, to be a good teacher one also has to be a master of the contents (subject knowledge, program and discipline) of teaching and related areas (BLM, CLN, CFC). A comprehensive mastery would enable him/her to elaborate the contents in an unambiguous way, to highlight the key points, to quickly provide proper examples, to show the connections with other courses, with experiments, with clinical operations or with applications (but he/she should pay attention to division of labour among courses and not go too far), so as to guarantee that the students have got the basic knowledge points. As CFC said,

The teachers who teach well are those who have mastered the course contents and related knowledge themselves. Otherwise it would be difficult to be precise and come up with timely examples, which often results in a boring course. CFC

To master the updated knowledge, the teacher should keep on learning – continuously reading related materials to enlarge his/her knowledge horizon (CFC).

- *Knowing how to teach and keep on improving*

Another attribute of a good teacher is knowing how to teach (e.g. how to organize the contents in a way that interests students and makes them easy to acquire), and being reflective on teaching – which aspects of the experience can be kept, and which aspects should be improved (BLM, CLN, CMM). Good teachers also keep on improving their teaching by learning from more experienced teachers, or by reading books on pedagogy and reflecting on their own experiences (BLM).

In addition, sense of humor and charisma, which attract students a lot in class teaching, are also regarded by teachers as important attributes of some good teachers (BCL, CZC, CMM, CWB).

- *Other preferred attributes*

BCG and BZK from University B and DGJ from University D mentioned students' evaluation of teachers, and stated that a teacher should not be constrained by the student evaluation of teaching or teachers. As BZK said,

Teachers should collect feedback from students, but should not try to please the students (e.g. give less assignments or easier exams) in order to get a better result in student evaluations. BZK

Some teachers (both young and experienced) (BCL, BLM, DCW) also talk about the difference between new teachers and experienced teachers. New teachers usually focus too much on the contents prepared in advance and their own lecturing and pay too little attention to students' response and interacting with them. In contrast, experienced teachers are able to detect the class dynamics and to mobilize students; they can come up with proper examples or cases very quickly, and adjust the contents flexibly according to the specific class dynamics, not just focus on finishing the prepared contents as inexperienced teachers often do.

The teaching staff's general attitude and behavior will also have significant educational effects (CLN, BCG). The teachers' characters possessing good traits may have an important impact on students, especially for them to internalize professional ethics. The students learn from how the teachers behave as much as, if not more than, they learn from what the teachers say. As BCG explained,

How the teachers behave naturally influences students' behaviors. BCG

Besides course teaching, teachers also have some unstated, informal, extra but sometimes expected responsibilities. Some teachers also mentioned that it is nice if a teacher can guide students in how to behave in different social situations, how to do things in general, and in choosing a career path either in or outside their profession (BCL, CZC, CZH). This is probably not stated in the official job description of teachers. These are not necessary components of teaching, but it would be nice if they were integrated into teaching.

Good teachers may be different and have different styles which are difficult to compare. In addition, good teaching may also be different, with different styles (DXY). Good teaching needs the students' cooperation, too (CWB). Ultimately, the final verdict of good teaching resides in society, with the performance of students in society (DZM, DZQ).

6.1.2.4 Summary of good teaching from the teaching staff perspective

From the findings reported above, achieving good teaching seems to involve at least three broad aspects, i.e. interaction with students, dealing with teaching contents, and teachers' personal attributes. With regard to interaction with students, the teaching staff advocated more class interaction, sharing relevant experiences and examples, keeping informal contact with students, and providing the students opportunities to learn by doing. As for the teaching contents, it is important to make the contents interesting and students interested, and also to give an overview. The preferred personal attributes for teachers are commitment to education, mastering of the teaching contents, knowing how to teach and keep on learning, etc.

To sum up, it seems to be very important for the teaching staff to learn more about the students, to relate to them both as practitioners of a certain profession and as more mature people in general, and to put teaching in the context of students' lives and thus let them make sense of the study in their life (e.g. to familiarize them with the contents by showing the connections with their past experiences, and making them see the usefulness or connections with the future).

6.1.3 Challenges, endeavors and expectation for support

This section reports on the challenges encountered by the teaching staff in their teaching, the endeavors they have been making and the support they expect in their work.

6.1.3.1 Challenges in teaching

The teachers talked about several main aspects that challenge them in teaching – their situation of being more encouraged to do research than teach by the university, big classes, increased teaching contents, and the unintended consequences of sticking to academic criteria.

- *The research focus of university and time-energy allocation of academics*

It seems that at all the three investigated universities the teachers are more encouraged to do research than to teach. Most of the teachers put more energy into research than teaching.

Among the indicators for assessing teachers, there are more indicators related to research than to teaching; and more weight has been put on research (BZG, BZK, CWB, CZC, CZH, DYL, DGJ, DZM). It is necessary, sometimes decisive, to do research in order to get promoted; and it is very difficult to get promoted by teaching, sometimes impossible for certain positions, e.g. a professorship. Research will bring the teachers both fame and fortune – they can get money from the research projects and use the research results, e.g. publications, to get promoted. Even though there are some teaching awards such as the National Teaching Achievement Award, National Excellent Course and National Excellent Teacher, it is far more difficult to get these awards than to publish papers:

If someone put all his/her energy into teaching, he/she would be, at best, an associate professor at the time of retirement. If, say, there were 10 criteria for promotion, 7 to 8 of them would be research-related. DYL

In response to the institution policy, most academics spent more of their time in research and publishing papers. Among 18 teacher interviewees who reported their work time distribution, 11 stated that they spent more time on research (BCG, BCL, BMN, BZX, CFC, CLN, CWB, CZH, DGJ, DYL, DZQ). Only one of them explicitly stated that he spent most of his time on teaching (BZG). Two of them reported that they spent more or less the same amount of time on research and teaching (BLM, DZM). Others claimed that it depends on the actual requirement or on which period it is of their work (CZC).

The most frequently mentioned factor that influences their time and energy distribution is the institutional emphasis on research in terms of evaluation of teachers or the university's policy, such as promotion of teachers, etc. Some teachers also mentioned that with the increase in teaching experiences there is a decrease in the amount of time needed for teaching preparation. In addition, it seemed that, since the results of research are more visible than those of teaching, it's easier to assess research by just counting the number of publications, while teaching is more difficult to assess. As one University C teacher said,

Everybody is able to teach and there are no absolute criteria for good or bad, so it is difficult to make [visible] breakthroughs in teaching. It doesn't matter so much whether it's good or just so-so. CZH

So, when a young or new teacher starts his/her work at a university, the first challenge encountered is research rather than teaching (CZH).

Others also mentioned that a lot of their time and energy has been spent on managerial work (e.g. BZK, who used to be a school vice dean at University B) or clinical work (e.g. the clinical doctor-teachers of University C).

Although the teachers are steered towards research, sometimes they themselves struggle to fulfill the duty of a teacher (DGJ) – trying to teach well under the university teacher assessment system, the university's policy and regulations, and trying to combine their teaching and research.

- *Big classes*

As mentioned above, class interaction is regarded as an important aspect of good teaching. The classes are often no less than 50 students, and some can be more than 100. Big classes make it difficult to raise and answer questions back and forth between the students and the teacher; and there is hardly any time for the students to present their work, express their ideas, or even get the opportunity to speak in class. Too many students also make it difficult, if not impossible, for teachers to guide all the students with their projects, and be close to students informally (BCL, CZC, DCW, DYL). Sometimes the teacher has to make an effort to speak loudly in order to make him or herself heard.

- *Increased contents with relatively insufficient course time*

There tend to be more and more courses with an increase of knowledge in each discipline, but the course time for some courses is reduced or condensed. This situation has meant that teachers are only able to outline the main points in class and cannot go deeply, into detail or demonstrate sufficient examples and cases, which often makes it harder for students to understand (BLM, CFC, CZC). Besides, it also restrains class interaction and the opportunity for students to do presentations or have discussions, and puts heavy burdens on the students in terms of reviewing all the contents, especially before the examinations. CWB described the situation at University C before the final examinations. He said that:

The students would have to stay very late reviewing the contents; and some students got very stressed and anxious. CWB

- **The quality dilemma: strict management, high academic criteria and failure rate**

When a university takes action for quality-enhancement, one of the approaches is to implement stricter management and stick to higher academic criteria. According to DYL, this causes serious problems for both the university and the teachers, and also the students. For the students it can lead to failure in exams, inability to get a diploma and disadvantage in job-hunting. As DYL explained,

Strictly implementing the academic criteria and the university policy means, if a student doesn't pass the exam, the teacher gives him/her a non-pass; and if the non-passes of a student rise to a certain number he/she will be flunked out according to the university policy. Also, the non-pass record can further undermine the students in job-hunting. DYL

For the teachers, the strict requirement can give them a bad reputation among students (for example, being called 'killer teacher') and make their course unpopular among students or lead to getting a low score in the students' assessment of teaching (BZK, DYL). The non-pass rate is also relevant in the assessment of teachers. There is an assumption that well-designed examinations will result in a normal distribution of the student scores; and the standard deviation will not exceed a certain amount. And:

If the non-pass rate of a course exceeds a certain percentage (for example 50%), it symbolizes either a low quality of teaching or the exam has been given in a too difficult way by the teacher.[9] DYL

For the university, the graduation (completion) rate and employment rate are among the references the Ministry of Education uses to evaluate the universities. Rising academic criteria will decrease both graduation (completion) rate and employment rate, and thus also be a symbol of low quality for the university.

---

[9] In China, most of the course exams are constructed and held by the teacher who teaches the course.

And sometimes this quality dilemma has been used by the students as a reason not to study hard, though not often. As one teacher said,

In fact if a student doesn't pass an exam, he/she gets to re-attend the course in the next semester; if he/she fails again, then he/she gets to attend the course the next and next .... This can go on until the eighth semester which is the final semester before graduation. The student may not experience so much pressure on them in this system. There is no limit for re-attending the courses and the ministry has stipulated that the university can't charge the students for re-attending the failed courses. When the final semester comes and the students are going to graduate, especially when the students get a job, we teachers can't be that wicked as to make life difficult for the students. Most of the teachers won't do it that way; we are, in principle, for the good of the students – 'now that you've got a rather good job, go back and have a good review of the textbook, and I will make the exam easier (so that you can get a pass and graduate with the diploma)'. It is quite difficult if the students are like this all the time – the seniors say this to the juniors: 'Don't worry, it will be fine in the eighth semester'; and the juniors tell their followers. DYL

- *The students' little interest in study*

Sometimes, students lack experience of actual practice and thus do not have a feeling of the significance of what they are learning (BLM, CFC, CZH). So the meaning of study for the students is to pass the examinations. DYL from University D admitted that getting high scores in the examinations of basic and program courses may not be very helpful in the students' job-hunting because they are not immediately relevant, or at least not sufficient. So the students may drift along the course with little interest, and then at the end of the semester spend two nights preparing for the examinations. That is all study means to them.

6.1.3.2 Teaching endeavors and experiments

There were no direct interview questions asked to the teachers about their endeavors and experiments, but during the interview many teachers talked about their trying to bring some new elements into teaching. This section mainly reports on how they have reacted to their current situation, especially to challenges and problems in teaching.

At University C, there is the institutionally-initiated adoption of problem-based learning (PBL) approach in teaching. At University B, the teachers are trying to bring new elements to teaching on their individual level. At University D, where all the teacher interviewees expressed their preference of student projects in teaching, the resources can only afford a small active proportion of students to get involved in projects.

- *Active learning at University B*

University B teachers were mainly trying to make the students learn more actively (BLM, BMN, BZG, BZX), such as by letting the student groups give lectures (BLM), group assignments (BMN), thinking training (BZG), and teaching through activities (BZX).

BLM from University B tried to let the student groups take turns to act out the role of course lecturer (teacher), with each group giving a lecture (or lectures) on one of the course chapters. This initiative is proposed by the students, and is a response to students' low tolerance to teachers' lecturing all the time and their willingness to participate more actively in teaching.

The students in charge of a certain lecture prepared the lecture carefully, and the class atmosphere was active. But the result was that the students only mastered well the chapter he/she had given lecture(s) on, and not very well chapters lectured by other students. The students lectured the contents as if they were reporting something, which is different from teaching. They might have prepared too much content, and not followed the student audience's understanding. And they lectured more or less at the same level as other students while I would have put it from and to a higher level. So some students had not listened to their peers' lectures so well and waited for the teacher to supplement. Sometimes, the teacher also tutored the student groups before they gave the lectures. BLM

The main problem with this kind of teaching is that there are too many students in a class, and there is often insufficient course time for all the students to do the lecture.

BMN has tried to give students group assignments and encouraged the students to interact more with peers and learn from each other, especially those good students to help the lag-behind ones.

BZG initiated a course targeted at training the students' ability to think. The course tried to confront narrow-mindedness and make students see the different possibilities or alternatives of some situations, for example, the different approaches to dealing with conflicts between people, which would facilitate the inner harmony of students, and then interpersonal harmony, and finally collective harmony. It mainly took the form of student group work reports, and class presentations.

BZX advocated teaching through student activities. She expected the students to participate actively in the course activities, and then to reflect on them. The textbook knowledge should make sense for students in and through their activities. But when she arranged activities, she felt that the students were somewhat confused – they were expecting the teacher to lecture them in the theoretical knowledge; they were not used to exploring for themselves and acquiring knowledge from their own experience.

- *Problem-based learning at University C*

In 2004, University C started an institutional initiation of introducing PBL in their teaching, which was enthusiastically supported and headed by the vice-president in charge of teaching and learning affairs. At the classroom level, the teachers implemented the PBL approach mainly by integrating case (disease or patient case) discussion in their classroom lecturing. Some teachers divided students into groups, asked them to find relevant knowledge related to the cases and then discuss it in class; and finally the teachers gave a summary of the knowledge discussed (CLN). Other teachers used cases as the starting point of a lecture by showing the symptoms of a disease or patient and raising some related inspiring questions to make the students interested in the lecture content that followed; and at the end of the lecture, the teacher returned to the cases and summarized the content (CMM).

When they use the PBL approach at University C, the teachers primarily use it as a good way to make the students interested in study and make the course content interesting. It doesn't change the teaching goal which is for the students to acquire knowledge. Elsewhere, the PBL approach was used to develop students' collaboration, communication, cooperation and management skills (e.g. Kolmos, Fink, & Krogh, 2004). This is obviously not the case for University C teachers, at least not in their conscious level. They mentioned that the positive effect of PBL is the increase in students' interest and curiosity in study (from studying for examinations to feeling like a doctor), and stated that different teaching approaches can be used to help students acquire knowledge. For example, CWB, a clinical teacher teaching regional anatomy, said, 'We conceived some disease case out of several related anatomical knowledge and showed it in class; when we were discussing a disease case, the goal was not to make the students know about the disease, but to make the students learn the anatomical knowledge through the disease case.' And CZH said, 'Only assured that they will master the basic knowledge can the students calm down to look at the cases.'

The challenge to use PBL for teachers, in some teachers' opinions, is that it requires the teachers to know more, to have more knowledge in related subjects, not only in the subject he/she is teaching. The PBL approach, where the students learn by looking for knowledge themselves and through discussion, might also destroy the knowledge structure from some teachers' point of view which assumed that the students only have limited learning ability. So normally the PBL element in teaching hasn't been given much course time. There is also the problem of dealing with too many students in one class.

- *Learning through projects at University D*

At University D, almost all the teachers were taking every chance to have more students involved in limited project opportunities. They thought that projects are the best opportunities for students to learn in an integrated way the engineering competences such as technical skills, manual skills, project management skills, teamwork, communication skills, etc.

6.1.3.3 Expectation of support for teaching

From the above-reported challenges faced by the teacher and endeavors they have made, some of the needed supports from policy and the institution have already emerged.

- *More weight on teaching*

Many teachers (BZG, BZK, CZC, CZH, DZM) proposed that the university should put more emphasis on teaching. The current situation is that research and publications have been given too much weight. When the university shifts some weight from research to teaching in terms of teacher evaluation, awards/incentives, conditions for promotion, etc., the teachers will be less pressed by research and have more time and energy for teaching.

In addition to the incentive change, the teaching staff also expected more course time to address the increased contents more thoroughly (BLM, CFC) and more opportunities for students to do projects (BLM, DZQ). As indicated above, many teachers emphasized the importance of students' learning by doing and engaging in real practice. So they proposed to arrange more opportunities (better infrastructure such as laboratories and equipment) for students to do projects, and internships (this may need external collaborations with schools for University B, and companies for University D, while University C has its own affiliated hospitals where a significant part of the curriculum is carried out).

For the young teaching staff, more opportunities for peer learning are also welcomed. As BCL proposed,

It would be nice to have opportunities to share experience and teaching materials (e.g. cases), and learn from each other among peers. BCL

In addition, the improvement of teachers' welfare was also mentioned by some University C teachers (CLN, CWB) as an approach to motivate teachers.

- *Physical condition improvement*

With the increase of students, there appeared to be some tensions in the use of some of the physical conditions. And some teachers (CLN, CZC, DCW, DZQ) proposed the improvement of physical conditions in terms of more classrooms to facilitate small-class teaching and discussion, better classrooms with multimedia equipment, better and more laboratories and experiment equipment, and more and better databases and information technology.

6.1.3.4 Opinions about teaching- and teacher-related awards

Not all teachers care about the awards given for teaching. One teacher, BCG, said that only those who have the prospect or potential to succeed care. For some teachers, probably most, it is not something within their reach, at least in the short term (BMN). Some of my interviewees said very explicitly that they don't care about any of the teaching awards and won't put any energy into them (BZG).

There is also the Matthew effect in the awarding process because the assessment of the candidates often refers to their previous awards (BZK) – if someone has Award A, he/she is also likely to get Award B, because having A is a symbol of his/her excellence which is considered in B's awarding process. If he/she has A and B, he/she is more likely to get Award C. And if someone has A, B and C, D is likely to follow easily. Reputation is very important in getting awards. As one teacher, who has been a member of an award assessment committee, said,

I have been involved in assessing and selecting the National Excellent Courses. I first looked into the members of the teaching team according to the indicators, especially the level/famousness of the team leader, not the details of teaching contents because I was not necessarily familiar with them. BZK

Another teacher, CLN, from University C, who was the teaching secretary for a course selected as National Excellent Course, described his experience of successful applying for the award which confirmed what BZK had said; he said that they had well-known professors in the course team, advanced laboratory equipment and teaching facilities, and provided comprehensive, detailed documents in the application, which were the key aspects of their application.

Besides, it seemed, in some teachers' opinions, that it was more difficult to acquire teaching awards than to publish. As DYL from University D said,

For many teachers, it seems that research will bring both fame and fortune – you can get money and you can use the research results to get promoted. Last semester our president initiated a policy to reward teachers more for teaching; for example, if you get a National Teaching Achievement Award with a prize of 20,000 yuan, the university will reward you with 10 times of that, i.e. 200,000 yuan. This can be of help to motivate teachers to put more energy into teaching. But these indicators for teaching such as the National Teaching Achievement Awards, National Excellent Course, and National Excellent Teacher, are far more difficult to get than publishing SCI papers, especially at our university (which is a provincial/local university, not so prestigious). There is only one National Excellent Course in the whole university, and the course holder told us that the course was not prepared when there was the National Excellent Course program; it was more than 20 years' effort. So it is not us teachers to blame. DYL

Setting awards is one important measure in the Chinese quality policy, as is seen in the *Quality Project* (cf. Section 2.3.2 in Chapter 2). In that project, one of the major measures is the National Excellent Course Program, and this program is award-equivalent in the sense that it is conferred a state of high honor with financial support. But awards seem to have only very limited impact according to the above teachers' reports.

6.1.3.5 Summary of teaching staff's challenges, endeavors and expectation of support

The main challenges facing the investigated teaching staff are the institutional focusing on research much more than on teaching, big classes, increased teaching contents, and the unintended consequences of sticking to academic criteria. The expectation of support is closely related to these challenges; the main institutional support expected is putting more emphasis on teaching and improving the infrastructure for teaching. The teaching staff are also making endeavors to improve teaching with or without institutional support, while teaching- and teacher-related awards seem to be of limited effect in supporting teaching.

6.1.4 Experience and opinions about the role of examinations in teaching

Examination in the form of paper-and-pencil test is the most frequently used method of assessing students' learning in Chinese universities.

6.1.4.1 The function of examinations

In the teaching staff's opinions, examinations have several different functions. Firstly, they are a promoter for the students to review what they have learned (BCL, BLM, CFC, CZH, DZM), so they consolidate learning. The students have to pass the exams to get credits and a diploma. Therefore they offer an external motivation for the students to learn. Sometimes they are also an affirmation of the students' efforts and give a student a feeling of achievement (DYL), and thus may further motivate him or her. Secondly, examinations are also a fair mechanism to allocate grades (BCL, CFC, CZH, DCW, DYL), so students can be differentiated fairly, for example in scholarship allocation or selecting students for further study. And thirdly, they are a relatively efficient method to assess how much the students have learned, though they may not be the perfect tool for assessment (CLN, DZM). As DZM said,

Many engineering courses are application-oriented; and they can be best assessed in application or practice. But exams can assess the students' study to some extent and in a relatively short time. In addition, exams are also a promoter and a binder for the students, otherwise there would be no one showing up in some of the courses. DZM

The function of formative feedback from examinations seems to be very limited (CLN, DYL), because it usually takes place at the end of the semester, and only provides the students with a score without comments or corrections. And examinations treat all the students on the same scale which cannot reflect the students' individuality (CZH).

6.1.4.2 The validity of examinations

Most teachers thought that the scores acquired by the students in examinations could only partly reflect their learning results (CFC, CLN, CZC, CZH, DCW, DXY, DYL, DZM). Usually the scores are consistent with a teacher's impression of a student during their interaction. The score is an important and necessary indicator of good learning and a good student, but not a sufficient one (CZC, CZH, DCW, DYL, DZM):

If the exam disciplines were well-kept, the results of the exams could be 80% consistent with the level of students' study – those ones who had listened carefully in class and who had asked pertinent questions would usually do well in exams. CFC

The good students would normally get good scores in examinations, but not necessarily the highest score. And high scores can also be acquired by rote learning (memorizing without understanding) (CZC). It is difficult to test a student's ability (e.g. manual ability, problem-solving ability) in examinations and therefore to demonstrate it by scores (DCW, DZM).

To sum up, in the teaching staff's views, examinations function as a promoter of learning, a mechanism of score allocation, and as a relatively efficient method of assessment. However, they offer only limited formative feedback. And the score of an examination is an insufficient indicator of learning outcome.

6.2 Students and learning

This section consists of the teaching staff's perceptions of what the students are and opinions about what the desired status of the students is.

6.2.1 The students' concern over employment

In the teachers' eyes, the students' main concern is their future development, especially employment after study, and most of them conduct their study accordingly (BCG, BLM, BCL, CLA, DCW, DGJ, DXY, DYL).

The students are concerned first and foremost about employment, their future job, and try to make sense of their current study in terms of its connection to a future job. The students study selectively on the courses provided by their programs based on the perceived relevance to their employment. They study extra contents which they think are helpful for them to get a (good) job and lack motivation for studying courses where they cannot see a connection with job-hunting and future work. When the students fail to see the significance, immediate relevance or usefulness of a course (in terms of employment), they find their study boring, or do not choose it at all if it is an optional course. And if they cannot see the significance of a required course, the meaning of study for them is reduced to getting a pass in the exam (CFC). DZM mentioned an attitude change among the students:

At the time they are taking the courses, many of the students do not realize the significance and importance of the courses for their future (and haven't studied hard enough). When they get to the job-hunting process and being interviewed or tested by a recruiter, the students tend to find the relevance of the courses and then they will study hard and actively. DZM

They also voluntarily study extra-curricular contents and take training courses outside the campus which they perceive as helpful for employment. They pursue certificates (English, computer skills, professional certificates, etc.) useful in job-hunting to ensure that their CVs stand out from the crowd (e.g. BZK, DYL).

If they have perceived a relative difficulty in getting a job as a bachelor (in fact, many of them do), the students will try hard to prepare for applying for further postgraduate study (preparing for the postgraduate entrance examination, and exams such as GRE[10] to apply for foreign universities, etc.) (CLA). And they will do their daily study accordingly, i.e. study selectively the contents relevant to the entrance examination (DGJ).

---

[10] GRE (the Graduate Record Examinations) is a standardized test that is an admissions requirement for many graduate schools in the United States, and some other English-speaking countries, or for English-taught graduate and business programs in some other countries.

The course exam result (and its rank in class) is an important part of both job and postgraduate study application, so the students care very much and can get very stressed by them (e.g. CLA, CWB).

Some students, influenced by the social atmosphere, sometimes yearn for a short-cut for success or quick success – to rely on the social capital or network of their family (BCG, CLA).

Some teachers also reported that there are some students who are confused with themselves; they don't know what they want or what to do with their study (BCL, CLA, BMN, DCW), and thus are easily influenced by others. Then what their seniors or peers do is an important reference for what they do. They can be motivated by witnessing their peers' or seniors' achievements or products, and do the study they are witnessing. They can also be easily dragged into computer games and things like that, and then leave their study aside. A teacher from University B, BCL, mentioned that:

A few students, who have sisters or brothers attending university before them, have got to know what university is about before they enter university; and they are clearer than others on what they can expect from university, and have taken actions in an earlier period in preparing for further graduate study, pursuing useful certificates in job-hunting or further study, looking for opportunities to study abroad, and things like that. BCL

Seen from the teaching staff's perception, their students have a strong awareness of social and societal pressure for employment and try to pursue employment-relevant contents and activities in their university study. They select courses according to their usefulness, choose supervisors who are application-oriented, pursue certificates useful in job-hunting (to ensure that their CVs stand out from the crowd), and go for further study in case of difficulty in getting a job. In the teachers' eyes, most of the students have an instrumental attitude towards study, i.e. study as a means of getting a job; only a very small proportion of students are interested in study itself.

6.2.2 The students' interest in study

The teachers attached great importance to the students' interest in their study, and some teachers categorized their students into three groups according to their interest (or willingness, activeness or voluntariness) in study (e.g. CFC, DCW, DZQ). The first group of students is the ones who have strong interests in studying their program and are learning actively and voluntarily.

The second group is those students who are uninterested in the program or discipline they enrolled in and make little, if any, effort in their study; and this group can be further divided into two sub-groups: the ones whose interests lie somewhere else, some of whom are pursuing a second program in their interested area, and the ones who don't know where their interests lie and don't know what to do.

Most of the students belong to the third group; these are the ones who are indifferent (neither interested nor uninterested) in their study because they are unclear about the significance of the study for their lives and future, and are learning somewhat passively (at least not very actively), just following the track and requirements of the university to get the diploma, and finally a job.

The lack of interest in study made the students inactive and unwilling to make an effort in their study. Many teachers reported students not initiating interaction with teachers, not raising questions in class (BMN, BZG, BZX, CZC, DCW). They may try hard to fulfill the study requirements and even get high scores in exams. But they are driven by extrinsic factors, not study itself. DYL from University D said that:

Some students made little effort in usual study, and then spent two whole nights just before the exams to get a pass. When it comes to the fourth or the final year of study, most of the students put most of the energy into job-hunting or preparation for graduate study application, and hardly any energy into their courses for that year. Thus some teachers don't like teaching the fourth-year students. DYL

The students should make sense of the course, see the significance or usefulness, and then they can show their interests. For example, the students are interested in the cell and molecule knowledge in the biology course when they can relate this knowledge to some disease:

I feel that from the undergraduate students' point of view, biochemistry is relatively difficult. The sense they made for this course is that it is a difficult course in examination. They cannot see much significance in this course for their future research or clinic work. In their eyes, it has a lot of knowledge contents with limited course time, and the contents cannot be thoroughly taught by the teacher and are hard to understand. So they have to study hard in order to pass the exam. Maybe after their graduation they can find the importance of this course. CFC

Some teachers believe that the students' lack of interest in study may be caused by their previous education (e.g. DCW).

6.2.3 Students' previous education

In some teaching staff's views, a lot of students (if not all of them) have been pushed to study by parents or teachers in their primary and secondary school periods; and they have also been guided in detail in study during those periods. And they get used to being pushed, being assigned, being organized, being guided in detail in study (BZG, DCW, DGJ). When they get to university, where there is no one pushing them as their parents and previous teachers did, they tend to distract themselves from study and/or don't know how to study because they haven't got into the habit of learning voluntarily, and haven't learned to learn by themselves. As one teacher in University D said,

Students can be pushed by parents or teachers in high school to get a good score in the higher education entrance examination in order to get access to university. When they get to the university, where there is no one pushing them, they lose themselves, because they haven't got into the habit of learning voluntarily, and haven't learnt to learn by themselves. When the teachers haven't taught this or that, they think it is not supposed to be learnt. When there is no assignment, there is no assignment. They are used to being pushed, being assigned and being organized. DCW

Another teacher at University B said,

Nowadays the motivation of the students is decreasing each year, so is their learning ability. Ten years ago the contents would be learnt by the students when the teachers had just mentioned several main points. But now we have to spoon-feed the students, tell them every detail. And this situation has much to do with the instilling manner of teaching in secondary and primary education. So they can't adjust to the way of study at university. The problem is not rooted in universities. All that the universities can do is trying to produce the best possible products based on the materials provided. The solution lies in innovation from secondary and primary schools, maybe from kindergarten. BZG

In previous education, seen from the teaching staff perspective, the parents and teachers have encouraged students to get a good score in examinations, especially the exams for the entrances for high school and for higher education in order to get access to further and better education opportunities. At university, the goal of pursuing further education is not as strong and clear as before; the students begin to think about their future – what kind of career/job they would like to go for, what kind of job would be available for them, whether or not to go for further study, etc.

6.2.4 Students' attitudes toward examinations

Most students care about examinations, and pursued a higher score in them (BLM, CFC, CLN, CZH, DCW, DYL). Sometimes when they cannot see the significance of courses in terms of connection with their future, examinations are how they make sense of these courses and the primary guide of their study:

For the undergraduate students the significance of biochemistry is to pass the exam and it is difficult, since they cannot make sense of this course in terms of future research or clinical work, which they haven't touched on yet. CFC

The students can more easily see the significance of examinations in terms of their effects on acquiring scholarships, further study applications, and job-hunting (CLN, CZH, DCW, DYL). Some students collected materials used in previous exams with the subjective aim of getting a pass with as little effort as possible (DZQ); this is a kind of study for examination.

6.2.5 Other issues concerned by the students

Besides the main concerns over employment, some of the students also have concerns about contests (especially for the engineering students) (DGJ), interaction with peers or collective life (CLA), students' clubs or associations, romantic relationships (CLA), part-time jobs (BCG, CLA), and playing computer games (BZG, BMN, DGJ). A few teachers also admitted that they only had touch with students in class, and didn't know much about the students' concerns (CLN).

6.2.6 Opinions about good students

In addition to asking the teaching staff what their students are like in study, I also asked them their opinions about what a good student looks like. And they mentioned the following characteristics of good students.

Firstly, good students have clear goals (BCL, CZH, CLA). They know why they are studying what they are studying; they are aware of the significance of the study in connection with their future, not just study for study's sake, and they can make choices in study accordingly.

I wish the students had their own goals, knew why they were studying medicine, and what kind of person and doctor they would like to be. Then they would be more active in learning. CZH

Secondly, good students are interested in learning (BZX, CFC, CMM, CZC, CZH, DXY). They are willing to learn, actively participate in teaching activities and learn voluntarily. There is no need to push them to learn:

The excellent students are usually active in asking questions and thinking, sometimes the questions can even push the teachers to move forward; they can relate what we are teaching to what have been previously learnt; and they can raise questions which relate to what we are going to teach; the questions they raised demonstrated that they have really thought about them. CFC

In addition, those students are not satisfied just by learning what is required; they try to learn even more, and come out with new methods or ways of thinking in problem-solving by themselves which haven't been introduced by teachers. And they voluntarily explore new opportunities or resources to learn.

The interest in learning can also be seen from the efforts they made in learning (CMM, CZH, DXY). Usually those students study diligently and persistently. And they can resist the temptation to have fun (e.g. computer games) now, and prioritize study. They give a lot of time and energy to study.

The interest and effort in learning, most of the time, gave them good scores in exams (BCL, CLN, CMM, CZC). Although exams don't necessarily reflect the ability or knowledge level, those students do well in exams. Exam score is important evidence of a student being good. But they don't have to be perfect in exams because then they would leave little time and energy for other activities (BCL).

Finally, good students are also good at arranging different aspects of their lives (BCL, CLN, CLA, DGJ, DXY). Those students are good at allocating their time and energy. They don't spend all of their time and energy on study; they also get involved in other activities such as part-time jobs, student associations, artistic activities, academic contests or research projects, etc., but they should schedule them well. Study is only one aspect of them. As BCL said,

Those good students should be good at energy allocation. I don't think that students should put all their energy into study. University is a place to develop one's potential. Maybe the students can spend about 60% of their energy on study, and then they should go and participate in some projects, join in student associations or clubs, take part-time jobs, organize some events, etc. The only thing is that they should schedule those activities well. BCL

6.2.7 Summary of teaching staff's perception and opinion of students

In the teaching staff's perceptions, the students were mainly concerned about their job prospects and tried to make sense of their study in terms of its relevance to jobs. And most of the students did not have a strong interest in study itself and were instrumental toward exams; and they had to be pushed to study, which, to a large extent, resulted from their previous education. In the teaching staff's opinions, good students are those ones who have clear goals during study, who are interested and make efforts in study, and achieve good results in exams. In addition, good students should also be good at arranging various activities during university study.

6.3 Institutions and higher education

In my interviews the teaching staff talked about their ideas about both the quality of universities and evaluation of higher education institutions.

6.3.1 Opinions about quality of universities

The teachers' direct opinions about the quality of a university can be divided into three main aspects, i.e. student-related quality, teacher-related quality and institution-related quality.

6.3.1.1 Student-related quality
Student-related quality includes both the study process and result.

In the study process, the high-quality universities provide their students with (1) abundant opportunities to learn, such as doing projects, and applying what they have learned in practice, especially those real-world situations outside the university (BMN, BLM, DGJ, DZQ); (2) a good learning and living environment such as an atmosphere where people love their study, good accommodation conditions, etc. (BCL, BZK, CWB); (3) greater freedom to conduct activities such as choosing courses (BCL); (4) good and responsible teachers (doing good research is one of the characteristics) (BLM, CFC):

A good university must have good teachers. I object very much to those teachers who say that the students are not good. For example, the students are too lazy to study; this is probably because they have not been enthused yet. I believe all students have good potential. BLM

As a result of their study, the students at high-quality universities should have (1) mastered knowledge, skills (both subject-related and communication skills), and their application in work after graduation (BCG, BMN, CWB, CZC, DCW); (2) got a good or satisfying job (high employment rate), found a proper career for themselves, or accessed further study (especially at more prestigious universities) (CLA, CZC, DGJ, DCW); (3) formulated a positive attitude towards life, and acquired some insights into thought and spiritual advancement (BCL, CLN):

How the graduates perform at work reflects the quality of a university. We would like the employer to recognize the work competence of graduates from our university and associate our graduates with good work competence. For example, the employer may employ our graduates expecting competence in experimental methods or high-level manual skills, and these expectations are actually fulfilled. CZC

6.3.1.2 Teacher-related quality

At a high quality university, the teachers should have good support in terms of (1) better infrastructure and physical conditions for teaching, such as classroom facilities (e.g. more classrooms to have smaller classes), laboratories, equipment, etc. (CMM, CZC, DYL); (2) a democratic and free atmosphere (BCL, BLM, BMN, CMM); (3) more opportunities to do research projects and actually do good research (CFC, CMM); (4) good work conditions such as supportive management, work teams, sufficient financial support, a good office environment, and more freedom to allocate their own time and energy (BCL, BLM,); (5) good life conditions such as higher salaries, etc. (BLM, CZC); (6) a good student source (to enroll better prepared students) (BLM):

A good university should provide the teaching staff with rich living and office conditions and a democratic atmosphere so as to prevent their energy being distracted from teaching. BLM

6.3.1.3 Institution-related quality

For universities as a whole, the good ones should also have (1) a good reputation (DYL, DCW); (2) good research production in terms of research projects and funds, as well as publications (CFC, CMM); (3) national awards, which enhance the reputation and put the university at a vantage point to apply for national funds (CMM):

We have got a 'National Teaching Team' and several key disciplines, which enabled us to apply for a lot of national funds and to recruit more well-known professors. CMM

6.3.2 Opinions about evaluation of quality

Some teachers also mentioned the evaluation of quality and stated how difficult it was to make evaluations (CZC, DZQ). In their opinions, it is difficult to evaluate universities because the impact of university education lies ultimately on how they perform at work. As DZQ put it,

I think the real quality of our teaching is demonstrated by our graduates' performance in companies, how they complete their work and the evaluations from the employer. DZQ

Quality of higher education may not be an issue within the higher education system; it may be influenced by the lower education system and other factors in society such as economic conditions. As BZG and DXY said,

Quality of higher education is not a problem of the higher education system and cannot be solved within it. It is a problem concerning both primary and secondary education (in terms of student source) and wider society (e.g. the influence of the family). BZG

The teaching aspect of a university is difficult to evaluate by employment rate, because employment rate is not only dependent on the university but also influenced by economic and other social/societal conditions, and employment rate is also a short-term goal which does not necessarily reflect students' career development. DXY

In addition, it is difficult to make comparisons between different institutions for at least two reasons. First, higher education institutions enroll differentiated student bodies; normally, better prepared students (in terms of higher scores in higher education entrance examinations) are enrolled in more prestigious institutions. As one University D teacher said,

It is difficult to identify the criteria for assessing the teaching quality of universities because universities enroll students with different qualities as the starting point of their education. Students with better entrance scores, who have a better starting point for university education, are enrolled to more famous universities. So it is of little value to compare universities in terms of teaching quality. DZM

Second, higher education institutions may have different focuses or missions (DXY) and are unevenly sponsored by the government (DYL). There are different types of universities with different focuses and priorities, and it's unfair to make comparisons without taking these factors into consideration.

6.3.3 Summary of teaching staff's views on university

In the teaching staff's views, a high-quality university should provide abundant learning opportunities and facilitation for students and rich support for teaching staff in both teaching and research, and achieve good results in these activities. It is difficult to evaluate the quality of higher education because the educational impacts lie on how the students perform in society after graduation. It is difficult to compare universities because they may have different focuses and they enroll students of different levels of preparation.

6.4 Summary of quality from the teaching staff point of view

From the above report of their opinions and perceptions of teaching and teachers, learning and students, higher education and university, it can be seen that teaching staff have the following main concerns over the quality of higher education:

- **Students' personal and professional development**

In the desired outcomes from teaching and university education, the teaching staff showed their concerns over students' development in terms of knowledge, skills, ability to transfer and learn, being clear about future development, employment etc. In teaching experience, the teaching staff reported their contextualized preferences for promoting student development by more interaction in class, sharing their own experiences, informal contacts with students, various strategies of dealing with teaching contents, and also in their teaching endeavors. The preferred personal attributes for teachers such as mastering the corresponding field, commitment to education, knowing how to teach, etc. also turned out to be closely related to student development. When talking about the quality of universities, they also mentioned student-related quality in terms of learning opportunities, learning atmosphere and responsible teachers, which are essential for students' development.

- **Teaching staff's teaching competence, research and promotion**

The emphasis on teaching competence can be seen from teaching staff's highlighting of the importance of knowing how to teach, mastering the class dynamic and teaching contents, the way teaching contents or materials are dealt with, endeavoring to explore new teaching methods and the opportunities for peer learning. The teaching staff's concern over their own research and promotion is mainly demonstrated in the challenges they faced in time-energy allocation, the university administration's disproportionate emphasis on research publication and national research projects, their expectations on policy and institutional support, and their ideas about the teacher-related and institution-related quality of universities. Performing good research can also be seen as a symbol of mastering the corresponding field.

- **Institutional infrastructure, policy and managerial support for teaching and research**

  Teaching staff's concern over institutional infrastructure, policy and managerial support for teaching and research is shown by their perception of the challenges they are facing with big classes, course time arrangement, time-energy allocation, tension between academic criteria and management requirements, and their support expectation in balancing the policy and institutional emphasis on teaching and research, in physical condition improvements, and the quality of the university in teacher- and institution-related aspects.

From the teaching staff's perceptions and opinions of their own work, the students and institutions, it can be seen that, on the one hand, they have devoted themselves to academic work such as teaching or the development of the students and research, but on the other hand, they have to complete their academic work under the framework of the university administration, from where they get support and resources. And the administrative focuses are not necessarily consistent with those of the academics, which has already emerged in my documenting of the teaching staff's challenges. So there is probably tension between them, which deserves further examination. This will be done in Chapter 8.

7

The student perspective

Students are the target group of higher education; their view on quality is also an essential part of the understanding of quality of higher education. In order to understand their view, I conducted 12 focus groups with 45 undergraduate students from three Chinese universities. This chapter first reports the findings from student focus group interviews around three broad topics, i.e. students and learning, teachers and teaching, university and higher education. Then this is followed by a summary of perceptions of quality of higher education from the student point of view.

7.1 Students and learning

The topic of students and learning is divided into four themes in the report below: (1) university and program selection, (2) learning experiences, (3) self-perception, and (4) opinions about exams.

7.1.1 University and program selection

An enquiry into how students selected the program and the institution they are now studying at is one way to learn about their expectations for university study. This section reports how the students chose their major or program and the university after the higher education entrance examination, and what factors they took into consideration.

7.1.1.1 Factors that influence university and program selection
The score they get in the National Higher Education Entrance Examination (cf. Chapter 2 for more information) is the basis on which students select institutions and programs for undergraduate study. Given a certain score, the selection would usually be a decision based on the job prospects of the study, the characteristics of the institutions and the student's study interest.

- *Score in the entrance examination*

A score in the *Gaokao*, the entrance examination, is usually a necessary condition for students to select an institution, where a higher score increases the probability for them to be admitted by a better (more prestigious and selective) institution. So before the *Gaokao*, the students, with the help of their high school teachers, put as much of their energy as possible into getting a high score. After the *Gaokao*, when they got a certain score, they would try to get into the best that the score allows them to.

In the interviews, many students (B1A, B2C, C2D, C2C, C3a, C3B, D1C, D2B, D2c) mentioned this consideration. One of them, a third-year girl student from University C, said explicitly that she wanted a better price-performance ratio (C2D). The price here is not so much the tuition they pay the institution for, but the score they have to get in the entrance exam in order to go to a certain institution and study their preferred program. She stated that:

I have chosen the institution with the best price-performance ratio, the institution which can provide me the best education based on my score. C2D

A second-year boy student said the score he had got constrained his choosing a better institution. He said,

I hadn't thought about studying here. But I got a bad score in the Gaokao and could not go to other better institutions. This is a relatively good one, so I chose it. I heard that this program had good job prospects. D1a

As already touched upon by this student, their second consideration is job prospect and career development.

- *Job prospects and career development*

Given a certain score they have got in the entrance examination, which is usually out of their control at the time of the selection, the most important factor taken into consideration by most of the students is the job prospects from studying a certain program at a certain institution (B3B, B3a, B4d, B4a, B4b, C1a, C2A, C2B, C3a, D1a, D1D, D2a).

They were very concerned about the employment situation of university graduates in China and expressed the prospects of getting a job when they graduate as an important factor in their selection. They are trying to select the programs or institutions that can provide them with the best job chances and professional reputation:

At that time, I knew little about the programs – only knew them by their titles. I thought all the manufacture tends to be automatized, and technicians in this area would be very popular. So I chose this. D2a

I watched some TV series before the Gaokao which depicted doctors in a very respectable way, and thought it's meaningful to be a doctor. So I am here. C1C

Some students from the rural area regarded study as an opportunity to improve the life condition of their family and move upward socially (B3B, B4b):

I hadn't thought about studying at University B; my parents persuaded me to choose it. They called me and said it's not easy to find a job, if you chose this there would be an assured job. I am from a rural area and have been studying hard for so many years; of course, I expect that I can repay my family and get a better chance to realize my potential. Most students in the teacher education program are not willing to work at the grassroots schools (in the small town). Most of our predecessors would have chosen to work at the key schools in the province capital or other big cities. B3B

In contrast to other factors, such as their own interest and campus conditions, that they had considered, job prospect appeared to be of more importance (B3B, C1a, C3a). A girl student from University B said that:

I had thought about my interest, but more important than that is the employment situation – the pressure to get a job is very heavy in China these days, so whether you can get a job (if you study the program) and the job prospects is the most important factor. B3B

A boy student from University C said that:

I knew the campus (physical environment and infrastructure) was not very good at this university, but the employment situation is quite good here. C3a

Some students were trying to find a balance between job prospect and interest (C2B).

- *Characteristics of institution and program*

The characteristics of an institution or a program are another important factor taken into consideration by students in their study decision. They are closely related to the job prospects of the study.

The first institutional characteristic to concern the students is its reputation (ranking, social recognition, employment prospects, etc.) (B3a, B4c, C2D, C3B, C3C, D2B, D3a, D3c). It appears that reputation itself, beside the educational experience it may represent, is an aspect of quality. It indicates a better chance of getting a good job for the students. Recruiters, frequently in a situation of dealing with many applications, often screen them by the reputation of the institution where they have studied. As one student from University B said,

When you go for recruitment, what the employer first learns about you is your diploma in your CV. Sometimes only a well-reputed university can enable you to get to the next step of the recruitment and to have the chance to show your competence. B3a

Besides the reputation of an institution as a whole, there is also the consideration of the reputation of specific programs within the institution where some programs have better reputations than others.

I wanted to be a teacher when I was in high school. And I knew that University B has the best teacher education program in China. So I had chosen this university before the Gaokao. B4c

Sometimes the program's reputation is more important than the institution's reputation. For example, a student from University C said that:

I put more weight on the program than the institution. Based on my score in the Gaokao, I would be able to go to (the more prestigious) University B; but if I went there I could only choose from its backward programs, which would lead me to a not-so-good future development. My parents disagreed with this situation very much. So I chose to be here. C3a

A student from University D mentioned similar considerations. She said that:

I considered two aspects, my interest and the job prospects. Actually, according to my score in the Gaokao I was able to study at South China Normal University or Guangdong University of Foreign Studies, which are better institutions. But if I went there I would not be able to study their outstanding programs. Besides, I like science and engineering, and I heard that the automation program here was very good, so I chose here. D1D

In most cases, it is better to have the opportunity to study somewhere in higher education than nowhere. But there are also many students who, if they can't study their preferred program at their preferred institution, would rather go back to repeat the last year of high school and participate in the higher education entrance examination again in the following year.

Some students also said that factors about faculty quality (C2D, C3C) and learning atmosphere (B1A, B3C, D1a) would have an impact on their institution selection. There should be famous professors or members of the Chinese Academy of Sciences and the Chinese Academy of Engineering among the faculty members. Learning atmosphere means that the students in the institution should love their study, and take study as their primary concern. Students get information about this mainly from public opinion, for example, mass media, talks with friends, high school teachers, relatives, classmates, etc. The situation of alumni, and their development now, was also taken into consideration by a few students (C2D).

Still another factor considered by many students is the location of the institution (B1A, B2C, B3a, B4a, D1b, D4b, D2a, D2c). In this aspect, the students have different preferences: some prefer universities near home; others would rather go to a different city to study. It appears that students from relatively well-developed areas such as Beijing and Guangdong would stay near home rather than go elsewhere (B1A, D1b, D2a, D2c, D4C). Students from rural areas preferred to go to big cities to study (and sometimes they have to because cities are usually where higher education institutions are located). A student (B4a) from Xinjiang, which is a developing area, stated that the students there would usually prefer to go elsewhere, especially big cities, and try to find opportunities for not coming back. Some of his classmates would select the city rather than the institution. He gave some examples of classmates who chose rather bad institutions just because they wanted to be in that city. A few students took location into consideration in relation to the climate, the convenience of interaction with their family, the living conditions in that area, meeting different people, etc.

Other characteristics about the institution considered by students include campus environment (B1A, C3A) and institutional history (B3C, C2D). The 100-year history of University B is regarded as an attractive factor.

- *Their own conditions*

In addition to the external factors, students also took their own conditions as factors influencing their educational choice.

The first one is their own interest. A few students mentioned that they had considered their own interest in their choice of institution and program (B3B, C1a, C1C, C2A, C2B, D1b, D1C, D2a, D2c, D4a). Mostly their interest had to be considered in parallel with job prospects (B3B, C1a, C2A, C2bB, D1D), the reputation of a certain program in a specific institution (D1D), and their score in the entrance exam (D1b, D2c). But sometimes their interest had to give way to the consideration of job prospects (B3B, C1a), as quoted above.

Another one is their family conditions (B3B, B4a, B4b), especially the economic conditions. The teacher education programs at University B are free of tuition fees, and some students chose them because they were trying to relieve the economic burden of their family. They also tried to use the opportunity of university study to extend their life chances. One of them said that:

The reason that I chose this university and the teacher education program (which doesn't charge a tuition fee) is the bad economic condition of my family. I came from Xinjiang Province. Usually the kids from there are eager to leap out through the Gaokao and university education to big cities and not come back again (to work and live in this backward area). A substantial number of students have been studying the teacher training program under the pressure from their families, not of their own accord. The reason is that the teacher education programs here are free of charge. Some of them might have been admitted to better institutions were their family condition not so bad. B4b

The third one is their perception of different expectations for boys and girls. There is the expectation that it's appropriate for girls to be teachers (B1D) and for boys to be engineers (B3a, C2C, D4C), which is also resonated, in students' or their parents' opinions, in the preferences of the employers:

I used to try my best to persuade my mother to let me study engineering. She said to me that the employers would prefer boys in engineering programs; if there were 10 people applying for a job, say nine girls and one boy, the employer would rather get the boy. C2C (a female student)

I had thought to study an engineering program because first I am a boy, then an engineering program is more likely to lead to jobs, and better life chances. B3a (a male student)

Still another condition about the students themselves is their physical condition. A boy from University B, B4a, said that he had thought about going to a university specialized in electric power to study electrical engineering, but because of his poor vision in his left eye he was not qualified for it.

- *Opinions of parents and some significant others*

When asked about their educational choice (the selection of institution and program), many students said they were significantly influenced by their parents (B1D, B3B, C1D, C2A, C2C, C3a, D2c, D4b), or other family members such as elder brother (B4b). Some parents had persuaded their children to put more weight on the job prospects (B3B, C2C, C3a), others just showed their expectation for their children to study a specific program (C1D) or to study for the same profession as they did (C2A); and still others had expected their children not to study too far away from home (D2b, D4b, D4C).

Some students also acquired information and suggestions from their high school teachers (B1c, B3a) or classmates (D1C).

7.1.1.2 Insufficiency of information in university and program selection

The students I interviewed also talked about the insufficiency of information about institutions and programs when they made their educational choice (B3B, B3a, B3D, B4b, B4c, D1C, D2a, D2B, D4b, D4C). Some students from the rural areas hardly had any chance to go online to check the information (B3D, B4c). Their choices may be based on very superficial information, for example just the title of a certain program (D2a).

Sometimes, even the students who had chosen certain programs knew little about them. Regretting their choice sometimes happens (C2B). One student from University B said that:

I wanted to be a teacher when I was in high school. I knew that University B has the best teacher education program in China. So I chose this university before the Gaokao. But I didn't know much about the programs. I chose educational technology, but I didn't know what it is about and what I can do with it. B4c

One student from University D shared her experience of having chosen a program which was not the one she had intended. She said that:

Actually I had intended to study mechanics. And then I saw two programs, 'Mechanical Manufacturing and Automation', and 'Automation'. I thought Automation must be more comprehensive, so I chose it. But now I have found that they are quite different, and even in different schools. D2B

Not only did the students themselves have little knowledge about higher education institutions and programs, their parents also had insufficient information (B3B, B3a, C2B). Lack of information may also contribute to students' focus on the external characteristics of institutions and programs such as online rankings (B3a).

7.1.1.3 Summary about university and program selection

In university and program selection, the main factors taken into consideration by the students are their scores in the entrance examination, the job prospects of the study, the institution and program characteristics, their interests, family conditions, gender roles, etc. They could be influenced by their families or significant others. In addition, they often made decisions with insufficient information. From the students' reports about their university and program selection, it can be seen that their main expectation for study at university is to get access to better job opportunities and thus expand their life chances.

7.1.2 Learning experience

In this part, I am going to report students' answers to my questions about their learning experience – not only about academic issues, but also examples that they have learned something in the broad sense, and that they have experienced a growth and personal development. First, I will present a summary of what they have perceived and regarded as learning and growth. Then the situations where that learning and growth have happened are presented, which at the same time will elaborate on that learning and growth.

7.1.2.1 Conceptions of learning and growth

During the interviews, the students reported various learning experiences, including academic and non-academic, knowledge and insights, personal and social. Those learning experiences involve different aspects of their learning and maturation as a result of a variety of situations and experiences. These aspects include knowledge, various skills, and personal change (see Table 7.1).

7.1.2.2 Situations of learning and growth

When asked about in which situations they had learnt the most, the students talked mostly about their experience outside the class, such as project work, application of knowledge in real-life situations, and social activities. And what they mentioned least was classroom teaching, which is their main form of study and where most time has been spent (cf. Chapter 2 for more information about the institutional setting of undergraduate study in China).

- *Project work*

Project work is contrasted with class lectures by the students. It is the learning situation most frequently mentioned by them – students from all the three investigated universities mentioned it. Five groups of them mentioned projects as their learn-most situations (B1, B2, B3, C2, D2). In three groups, the students unanimously agreed on the benefits and positive experience of project work (B2, B3, D2).

Project work is usually a task given by the teacher as part of a specific course. It is often aimed at students' applying the course knowledge. Students can do projects in groups or individually. Besides being part of a certain course mentioned mainly by University B students, projects can also be organized around certain contests, as is the case at University D (D4b). Or some students may have the opportunity to be a research assistant (C1B) or project member in their teacher's research project, as at both Universities C and D.

Table 7.1 Students' perceptions of learning and growth

Perceived learning and growth		Example
Knowledge	Subject knowledge	Calculus, physiology, microcomputer theory
	Generic knowledge	Knowledge about professions in society, different cultures, knowledge that widens personal horizon; there will be gains with efforts in spite of bad circumstances
Skills	Technical skills	Computer skills such as Photoshop, Word, programming, technical drafting Clinical skills such as diagnosis (body check), surgery Manual skills such as making a printed circuit board
	Practical skills	Make plans, schedule, prioritize issues, deal with specific situations
	Cognitive skills	Comprehensive, precise and logical thinking, clear expression of ideas in presentations, the logic of diagnosis, different perspectives in thinking about the same issue, know what and how to learn in future
	Social skills	Teamwork, division of labor and cooperation, tactics of interacting with people of different characters (peers, patients), negotiation (with sponsors), organizing events (performances, volunteering activities, etc.)
Personal change	Character	More confident, open-minded, patient and composed
	Identity	Personal (gaining friendship), academic and professional (tacit rules in hospitals)
	Self-awareness[11]	Awareness of maturation and responsibility, knowledge about oneself, less self-centered, clearer about career development and life prospects

In projects, students can get the most gain from their point of view – all the aspects identified above, i.e. knowledge, skills and personal change. They reported many positive experiences in doing projects.

First, the students experienced projects as good motivating starting points for learning and as opportunities to make rewarding efforts where there were showable products.

The project task or problem may serve as the impetus and guidance for the students to organize their learning. It offers the students a point of

---

[11] This was reported by some students (B2a, B2C, B4a, C3C, D4C) but not in association with any specific learning situations.

departure and certain kind of reference to which they can organize and adjust the learning activities. As B3D put it,

I think the situation where you can learn most is like this: when you have something to do, and you know what to do, you have a center to go around, for example, collect data, you have a criterion or benchmark to radiate, then you have combined your own thinking and doing. B3D

This function is also illustrated by C2D's experience:

In our first year, we were all asked to write a paper in the physics course. For almost all students it was just a burdensome task. But I was quite interested in the heart at that time. So I went to the library and read a lot of books. And finally I wrote a paper on the thermophysics of cardiopulmonary bypass. It's a very tiny aspect [of this area], but I had read a lot about the heart. Then I felt fulfilled and quite happy, because this was what I wanted to know, to explore. And actually I got to know a lot. C2D

Compared to class lectures, doing projects also provides students with more autonomy (they have more impact on the outcome) in the learning process, so they can make their own efforts more actively. Several students from University B shared their experience in this aspect:

In contrast to class lectures where the teachers teach what they would like us to know, during projects we get to know what we should learn in order to solve the problem, and then we learn it ourselves. It is more impressive to learn by doing. B2b

With every group assignment comes the presentation where you can feel the achievement, because before the presentation, you have to have gone through a process of project design and implementation, and during this process the group members must have tried their best in cooperation and communicated with one another their sparks of wisdom. I don't think we can learn so much as assumed in class. B3B

It seems that students' feeling of achievement and satisfaction intensifies when more efforts have been made in projects. As B3a said, 'When you get a difficult task, you spend a lot of time on it. Then when it is completed, you will feel that you have learned a lot.' And this feeling is sometimes stronger with the teacher's recognition and praise for the work; but the students are not dependent on teachers' praise, as the following group discussion illustrates.

Interviewer: *Have you experienced in the study process some moments of achievement and satisfaction?*
B1D: *Yes. It's when we have completed our group project – made a long-time effort and finally we get our result.*
B1B: *And then do a presentation reporting our project in class.*
B1D: *Then we feel proud.*
Interviewer: *Can you describe which factors might enhance this feeling? For example, the difficulty of the task.*
B1D: *When we have done a good job and the teacher praises us, the feeling of pride can be stronger. When we finish our presentation, the teacher says "well done" and encourages us. Then all members of the group feel very proud.*
B1B: *It can also be a situation where all members of the group have made a great effort and everyone has done a good job. Even though we haven't got any praise from the teacher, we could also feel contentment.*

Second, the students experienced projects as opportunities to unite knowledge and practice, to combine their thinking and doing. To them, the class-taught theoretical knowledge is abstract and distant from real life. Projects offered them opportunities to put the knowledge into use with their own hands and thinking; and therefore, it became tangible and animated. As illustrated by the following discussion:

Interviewer: *When do you think you have learned the most in study-related activities?*
D2B: *The project part of courses and laboratory work where you can apply the theoretical knowledge in class and make some real objects. Sometimes you are taught a lot of theories in class, but you don't know how to apply them. ... For example, the (programming) course only tells you what the small functions of specific sentences are. But when you design and make some real gadgets, you have to think for yourself about how to put many sentences together to materialize certain bigger functions.*
D2a: *I have quite similar feelings. For example, it seems there is nothing easier than the connection of a lamp circuit, with only one bulb and one switch. But when you actually do it, it appears to be not as easy as you had thought.*

D2c: *There is some distance and discrepancy between theory and reality.*
Interviewer: *What's the difference?*
D2c: *Your manual skills. The theoretical principle, and one bulb, one switch, then you connect ...*
D2a: *In the circuit chart we use one line to represent the electricity wire. But in reality there are three power wires – the live wire, the null line and the earth wire. That's the difference.*
Interviewer: *So the theories are a simplification and abstraction of reality?*
D2a: *They are simplifications.*
D2c: *And abstract.*
D2a: *So you should use your hands more and touch reality.*
D2c: *Yes, use hands more often.*

In the process of doing projects they not only consolidated and internalized the knowledge (B3C), but also gained practical wisdom such as in project management and team work (B2a):

It is practical – we can do it ourselves using our own hands, and we can experience the process – project management, the interaction and cooperation of group members. B2a

Third, the students also took projects as opportunities to develop meaningful relations with peers and gain friendship. In group-conducted projects, the students have to interact and collaborate with each other to finish a shared task. During the process, each student is expected to demonstrate their merits and contribute their part to the project. This simultaneously provided them with opportunities to learn about one another's merits and character, and to develop friendship. As the following discussion shows:

B1A: *That may be some kind of friendship – we have recognized each other, or we have seen our peers' merits, or we get praised by a peer. Then I feel very good.*
Interviewer: *Does this happen very often?*
B1A: *Yes. In the collaboration process you get to know your peers' merits which were unknown before. I think it quite meaningful, no matter what result you get.*
...
B1c: *We have more classes than group work. The opportunities for collaboration are too few. We are individuals in class.*

But the positive effects of projects also depend on certain conditions. The most important condition seems to be the teacher (or supervisor. The teacher teaching the course is the supervisor by default and supervises all the students taking the course). The optimum situation is that the teacher shows great care (B2a) and attention (B2C), has strict requirements (B2C) and high expectations (B2a) for the project, lectures less leaving more time for projects (B2a), and gives frequent feedback (B2b) and opportunities for class presentation and discussions with both the teacher and other groups (B2a), and probably also praise (B1D, B2a):

Only some of the projects are productive, not all of them. It depends on the group and the supervision of the teacher. If the teacher makes strict requirements of students and gives a lot of feedback, it can be productive. But if the teacher is just inattentive, and the group members are not active, it can be just one person writing a paper. B2C

Then the projects should be of proper difficulty and feasible for the students to complete with effort (not exceeding their ability) (B2b). The course should be allocated enough course time to arrange group assignments (B2a).

In addition, the students were also aware of the limitations of projects. They may not suit all courses, but are more suitable for the practical or operational courses where the students can produce something (B2a).

With regard to the relations between knowledge taught in class lectures and projects, the students once more mentioned the role of teachers. If the teacher requires a direct application of the theories taught in class or has taught the knowledge in a detailed way, the students will follow him/her more. Otherwise the students will become more divergent and explore more by themselves. The following discussion illustrates this.

Interviewer: *What do you think about the relation between the knowledge taught in class and the projects you have done? Can you use the knowledge in projects?*
B3B: *I think it depends on the subjects (courses) and the teacher's requirements. Todays' students, we are a bit opportunistic (would try to get through using short cuts) – we would have asked our seniors about the styles of different teachers; if the teacher tends to require students to fully apply the theories taught in class, we tend to do so; if not, we tend to use the theories taught by teachers in a broad sense and go in more divergent directions according to our own interests.*

B3C: *I have some different experiences. When I do a project in a course where the teacher has taught in a very detailed way, I tend to do it step by step with the approach taught by the teacher. But in some other courses where the teacher has only taught the basics, then you have to complement with extra knowledge in your course project, otherwise it's impossible to complete it. So it depends on the project requirements and the knowledge offered by teachers. You have to adjust according to the needs of the project – if there is a need to supplement, then you supplement; if the teacher has taught in a very detailed way and there is no need to supplement, then you just do as the teacher has told you.*

It seems that projects, for the students, are not only good activators of learning, but also a popular way of organizing the process of learning. Doing a project is a process combining reviving the learnt knowledge, thinking and doing. Students learn in the direction of the project they are doing, thus they can learn and at the same time see the process and result of learning – a project report or something they have designed. This gives them a feeling of achievement and satisfaction. When doing project work, they can explore new knowledge themselves and put the already learnt knowledge into use, thus acquiring a deeper understanding. They learn together with peers, cooperate, learn from each other, get feedback and recognition from each other, thus not only learn some knowledge but also learn about people and how to cooperate with each other.

- *Internship and application in real-life situations*

While projects are more highlighted in their learning experience by students of engineering or engineering-related programs from Universities B and D,[12] the medical students from University C emphasized a lot the importance of internship and application in real-life situations (C1E, C2A, C2B, C2C, C2D, C3B). Touching on real problems (people's diseases or confusions) and knowing about their professional practice (hospital situations), in contrast to textbook knowledge and laboratory work, were repeatedly mentioned in their learning experience by the medical students.

---

[12] The student participants from University D are from engineering programs, i.e. automation, electronic information science and technology, and electrical engineering and its automation; and the student participants from University B are from the educational technology program, which is an interdisciplinary program with engineering elements.

It seems very important for them to know the real-life relevance of the knowledge they are studying.

The real-life problem may also serve as a motivating starting point for the students to learn, or as an opportunity to put knowledge into use. This learning not only results in knowledge acquisition but also the mastering of approaches to acquire knowledge, since active thinking is involved:

C1E: *The situations where I felt fulfillment or achievement are usually the ones when I encountered some medical problems in everyday life. In those situations I would try to use my knowledge to get an answer to those problems. The answers are not necessarily correct. But they are acquired through my thinking. I think it's a process of establishing my medical thinking. So I feel quite good.*
Interviewer: *Can you give some specific examples?*
C1E: *For example, some relatives may have some pains and they consult me about what could be going on with their body. And I might help them analyze the situation; maybe a muscle has been pulled during physical labor, or maybe there is some inflammation somewhere. And other such small problems.*

During the interaction with people who come with problems or confusion, the medical students are also gradually establishing their professional identity. As illustrated by C3B's experience:

When there is some holiday and I go home, I am usually taken as a professional expert by people around. Although I have very limited knowledge about medicine, they take me as a doctor. When they get pains somewhere on their body, they come to consult me about what could be the matter. Or when they get their diagnosis document from the hospital, they come to me and ask me to explain what the terms mean. And I can also help them calm down. Although I am still a nobody in this profession, I have a feeling of achievement. C3B

Having a taste of professional practice during internship at the hospital (maybe just a few days) can also serve as a motivating starting point for the students to learn, and make what they have learnt more meaningful to them. This taste helps them to get acquainted with their profession and construct their own vision as a professional practitioner – what knowledge they have to acquire, how knowledge is used in practice, how to interact with different patients, and what the professional culture is like. It is in this process that the students get to know and feel what a doctor really does and looks like. As illustrated in the following discussion:

C2B: *I felt satisfied recently because during this winter holiday I spent four days in a hospital as a kind of internship. For some classmates it was perfunctory; and others did not even appear. But for me I learned a lot about our profession. It is not some specific knowledge; just experiencing the atmosphere of this profession makes me satisfied. I went into the real situation of this profession in real life. In that process I knew what I should do in future.*
C2C: *No longer just learn from the books. I had a similar feeling when I went to the surgical oncology department in the hospital this winter holiday. Actually until now I knew very little about it, and I haven't learned diagnosis. But the patients made me feel that I was trusted as a doctor. Just their eye expression could make me content.*
C2D: *My experience was a bit different. The communication between doctors and patients was quite interesting. Usually the doctors would shout at patients. If you don't shout at them, they seem to look down upon you. Only those veteran doctors are respected by patients. The trainees like us are not. ... The patients I encountered were those with endocrine dyscrasia. It's kind of chronic but not serious; they have to take pills every day. Some of the patients may have economic pressures. So they might have a bad attitude. And we have to pay attention to the tactics in interacting with different patients.*
C2B: *Another satisfactory aspect is that when you go to clinical situations you will find out a lot of things you don't know. Then you come back to study and get to know the issues you have encountered. That's a very good feeling. And you will have the knowledge well knit, which is much better than just listening to course lectures.*

- *Class lectures and teacher's influence*

Although some students mentioned the teacher's influence in their learning experience (B3C), very few of them mentioned class lectures. When I asked about their experience of class lectures, they regarded them as a passive routine with no feeling of fulfillment or achievement (B1D, C2C). The teachers give the lectures (B1D), and the students listen and take notes, seldom actively doing anything.

7 The student perspective

Interviewer: *You have talked a lot about your learning experience outside the classroom. Then what about the classroom teaching? Have you ever experienced learning a lot in class?*
C2A: *I think some teachers are quite good. They give us the slides before the class, and explain the key points in a very clear way, sometimes a bit more than the textbook contents, sometimes a bit less. They are quite experienced and know this profession quite well, and based on their experience they highlight the important contents, so that we don't have to go through the whole textbook, which is hardly possible. They teach us in a well-organized way and we gain a lot.*
C2C: *I don't mean to belittle teachers' efforts. But the knowledge I get (from class) is quite banal.*
Interviewer: *With no deep impression or ...?*
C2C: *We are impressed. But the whole process brings no feeling of fulfillment and achievement. It's just you teach and I learn. We began this since we were little kids.*
...

C2C: *Maybe there is some gradual impact that potentially influences us. But we don't know.*

Sometimes the students cannot see the usefulness of contents (theoretical knowledge) taught in class lectures and lack the motivation to learn (B1B); although they have learnt the subject knowledge, the only relevance left for them is the exam. So they found that theories are boring (B1D), abstract, difficult to absorb (B1A) and easily forgotten (B1c, B1D). Also, the theory classes are usually taken by a lot of people in a big classrooms (B1c).

The only exception regarding class lectures is when they are about technology, which the students can put into use and do something with outside class:

Interviewer: *When do you think you have learnt the most at the university? In what kinds of situation?*
B1B: *Technology class, like computer.*
B1c: *Like technology experiments, I feel I learned a lot. The theories you learnt ...*
B1D: *Yes, when you learnt then you forget.*
B1B: *And we don't want to learn.*
B1c: *When you have learnt it, you forget it. But with technology, you may use it sometimes.*

7.1 Students and learning

Interviewer: *Can you say more about that?*
B1D: *Take the course about Photoshop and web page design as an example. When you work at the student union, there may be some journals or newsletters to produce. Then you can you use the computer technology you have learnt. You can use it outside class. But the theoretical knowledge is quite boring.*
B1B: *And there are usually a lot of people in the theoretical class.*
B1D: *I feel it is useless. Maybe it is because I haven't seen its usefulness. Maybe there can be some.*

Although the effect of class lectures is quite limited in the students' reported learning experience, the students are clear about the influences of some impressive teachers. Although those teachers are impressive in various ways, they are all trying to relate themselves or what they teach to the students, none of them being impressive by just teaching subject knowledge, which is their official/formal responsibility (they are impressive as a person, not just a teacher qua teacher). Those various ways of being impressive include (B3C, C1D, D4C):

- Sharing with the students one's own learning experience,
- Putting the teaching contents in a broader context or relating them to recent advancements in the area (which is experienced by students as making them feel they have a lot to learn),
- Showing the usefulness of the theories taught (for example, one micro-economy teacher encourages the students to analyze everyday phenomena such as the change of house prices in China with the facilitation of theories, then the students can make their own judgments so as not to be easily cheated),
- Commenting on the issues of concern to the students,
- Showing personal style and charisma, for example, being precise and logical, and being affable (those personal traits not only influence the students as developing persons but also serve as a model for that profession, as B3C said about her teacher: 'His teaching style will have an impact on my teaching style when I become a teacher.').

And the impressed influence of teachers highlighted by the students lies not in knowledge, but in cognitive, social and practical skills, and personal changes.

The students also complained about their too few opportunities to interact with teachers (B4a, B4b, C3C). Usually the teachers are only met in class giving lectures, and:

Usually when the class is over, teachers go back to their offices if students do not ask them questions. So we have few opportunities to interact with them. C3C

- *Social and non-academic activities*

Besides study-related activities, students also reported a lot of non-academic activities in their learning experience. Non-academic activities is used here as an umbrella term which covers all activities that are not required in the formal program curricula or related to the subject matter. As mentioned by the students, it includes activities organized by student organizations (student association, clubs and student union), voluntary work, part-time jobs, sports, etc. Almost all of them are social activities where the students interact with all kinds of people and encounter different situations they have to respond to.

These activities are usually initiated and organized by students themselves with no teachers guiding them. The teachers can be consulted sometimes, but the students have to organize activities themselves. So they learn how to organize events by themselves, how to work as a team and cooperate. By doing these activities themselves, they gain experience in making efforts and become more confident in themselves. In addition, they also learn about people's traits, and gain experience in interacting with people of different characters. Last but not least, they make friends.

Life at university is more diverse than at high school, especially in terms of social activities. By experiencing various activities, the students first learn practical skills such as how to schedule and prioritize affairs and how to deal with pressure (B4a, B4b, D2a). As illustrated by the following two examples:

I am the monitor of our class which I haven't been before. At the beginning, I was very quick-tempered and there were a lot of issues to deal with. So I felt quite stressed. Then gradually I learned to prioritize the issues – first I deal with the most important ones, then the second most important ones, ..., leaving the unimportant ones undone if I don't have the time and energy. Now I have quite some experience in this. Now I have also become more patient than I was. B4a

I am in the student union and in charge of contacts with different sponsors. Holding events always needs some financial support. And it's my job to raise funds. This used to exert a lot of pressure on me because my negotiation with the sponsors determines whether or not we can get the support. Now I have a lot of experience in this, and still I have pressure with this job, but I know if you give yourself too much pressure, it is counterproductive; you have to control the pressure and make it into impetus. You have to care about it, but you shouldn't care too much. And sometimes you have to let it go, or let the enemy off in order to catch him later. D2a

By performing certain social roles or tasks such as organizing student association events, being a volunteer guide in the museum, being an Olympic volunteer, doing part-time jobs, etc., the students have gained real-life experience of interacting with people. In these processes they have gradually gained the social skills of how to proceed and at the same time overcome shyness and timidity, and become confident in themselves (B1A, B2b, B3B, B3D, B4b).

At first you may think that it's impossible for several students to organize such a big event. But when you get down to doing it step by step, eventually you succeed. This experience might make you feel that things may not be as difficult as you originally imagined; but you have to make an effort in order to succeed. B1A

In the social activities, the students gain more experience of interacting with different people and understand more about interpersonal relations (B4a, C2C, C3B, C3C, D2a, D3a, D4C). Thus, they become more skillful in communication and negotiation, and adjusting oneself to situations. For example, D4C from University D said that 'in high school we used to quarrel with classmates in the same dormitory, but now we never do this'. Not because there are no conflicts any more, but because they have learned how to deal with those conflict situations. Another example is C2C's experience:

I have worked as an assistant in the office for student affairs. This is the first time I have interacted with a teacher[13] as workmates. And they trust

---

[13] The administrative staff are often also called teachers in China.

me a lot, although my experience of this kind of work is very limited. At first we didn't understand each other very well, but they tolerated me. Then gradually we saw each other's good-heartedness and got on well with each other. They have helped me with practical issues and taught me tactics. When you feel their trust, you will also show kindness. Although they haven't taught me any knowledge about medicine, they taught me how to interact with people and how to arrange events. I feel quite grateful. C2C

In the experience of interacting with different people, the students also become more tolerant, open-minded and patient (B4a, D4a):

D4a: *In my first year I knew little about university and just focused on my study. Then I joined the university troupe and spent a lot of time there, leaving only one-tenth of my time for reading (studying). It resulted in being very busy (preparation for exams) at the end of the semester. But since that year my character has changed a lot – I have become more open-minded.*
Interviewer: *Can you give some examples?*
D4a: *Take organizing an event, for example. Usually there would be about 10 people in charge of organizing an event. But it often turns out that only two people are actually doing it. Others may just come to have a look and then go away. Or they may just make some comments or cynical remarks. I usually take it as an opportunity; they don't grasp the opportunity and leave it to me, why not just take it. Actually it can be organized by three or four people, just that we have to do a bit more work and get busier, but it's no problem for me.*

- *Other situations of learning*

In contrast to the specific situations of learning discussed above, self-study, mentioned by a few students, does not refer to any one specific learning activity; instead it highlights the main source of motivation for study – the students' own interest, and it is self-initiated. When the students find their interest, they use different resources such as relevant lectures, library, etc. to explore and learn by themselves (B3C, D1b).

In addition, a few students also reported in their learning experience situations such as preparing for the exams where there is something pushing them to learn (C2), participating in interesting lectures which widen their horizon (B1), and doing sports (C3a, D3b) where they can be trained to be more composed in matches.

7.1.2.3 Summary of students' learning experience

From the above report about their learning experience, it can be seen that the students have experienced a broad range of learning and personal development, such as acquiring both subject and generic knowledge, mastering all kinds of skills from technical, practical and cognitive to social skills, and personal change in terms of character and identity formation, and improved self-awareness. The most highlighted learning opportunities are project work, application in real-life situations, and social activities.

From their reports we can see that the students, during their university study period, have a broad concern over their personal development and growth, not just the subject knowledge that they are studying. In these learning experiences there is a strong social dimension, a real-life relevance or usefulness dimension, and an activeness dimension. The three most highlighted learning situations are all ones that enable the students to learn in a community and in interaction with other people. This not only motivates them to learn, and makes learning meaningful (cooperation, mutual recognition, fulfilled identity, relatedness, etc.), but also provides opportunities for them to learn from each other. The real-life relevance or usefulness dimension reflects the students' concern over what they can do with knowledge in their study: either it can facilitate producing something as in projects, or it can solve or explain people's problems as in medical applications, or it can inspire students how to proceed on social occasions as in the social activities; it is not knowledge for knowledge's sake. The activeness dimension refers to the students' interest or willingness, and autonomy in the learning process. In all the highlighted situations, the students are all highly motivated and actively engaged, and they have, at least to some extent, the discretion to take different measures and thus have an impact on the outcomes.

Project work is relatively highly scored in all these three dimensions and thus the most popular learning situation among the students, while class lecture, where the students learn individually and often passively, and sometimes deal with theoretical knowledge with unclear real-life relevance, is the learning situation least mentioned by the students. In addition, the students' suggestions to make class lectures more impressive also point to these three dimensions.

7.1.3 Self-perception

This self-perception section reports my findings on how the students perceive themselves in their main concerns and activities during university study, and what their idea of a good student is.

7.1.3.1 Main concerns

Generally the students are most concerned about their future which is most likely to be in one of two paths – going on to postgraduate study (B2C, B2, D1) or going directly to work (B1, C4, D1b, D1C, D2). Going to postgraduate study and pursuing a higher degree seems to be just a means to get a better job. So a job is the ultimate concern of the students. Most students have been trying to choose activities relevant for jobs and make sense of courses in relation to jobs.

During their study, the students usually also keep an eye on their job prospects, which includes aspects such as job-related policies (e.g. the policy of the teacher training program at University B[14]) (B1), job fair information (the positions on offer and corresponding requirements) (B1D), the job-hunting situation of their seniors (B1, B3B, D1), and expected work conditions and status given the diploma they get (B1).

When they have perceived that their undergraduate diploma is not enough for the acquisition of expected jobs/positions, most students (and more and more students) deem that they would better go further on to postgraduate study and equip themselves with a higher degree (B2, D1B, D1C). A higher degree is expected to be associated with a better job (D1B). Many students give a lot of energy to preparing for the entrance examination for postgraduate study. Even when the graduate study is not their interest, they would still pursue it, as mentioned by B2C about the experience of one of her schoolmates:

I have a schoolmate who is excellent and has been admitted to postgraduate study with exemption from the entrance examination. Then he joined the lab work (of postgraduate students). But he is very uncomfortable with it; he thinks he should have gone out to work. The problem is if he leaves the campus now with his undergraduate diploma, he

---

[14] A policy requires a contract between the students, the institution and the educational authority of the student's home province which offers the student free undergraduate education in the teacher training programs and requires the student to go back to work at schools in his/her home province after finishing the study.

is hardly likely to get a good job. This forces him to do the research work as a postgraduate student. B2C

Some students also care about the assessment result in terms of both score in the exams and other sources of credits (C1, C3) such as publishing papers, participating in contests, and activeness in school-organized activities, etc. Scholarships will be awarded based on the assessment. To the students, this may also be important both for job-hunting, as it makes a better looking CV (D1), and for postgraduate study, as it is one of the criteria for the exemption of the entrance examination (B2), and also for choosing one's specialization for the medical students (those ones with better assessment results have the priority).

The students hoped that the university career guidance and counseling would start from the beginning of university study (B2), so that they could be clear at the beginning on whether to go further on to postgraduate study and thus an academic career or to get some specific (outstanding) competences which enable them to do certain work/jobs immediately (B2b).

These concerns also have impacts on their study. As students from University B said,

When there is fierce competition, you must study hard to stand out. B4b, B1D

Feeling there is not much hope for the future, they get a bit gloomy and passive in their study. B4c

7.1.3.2 Activities
When asked what their main activities are, the students reported a variety of activities. These activities can be divided into three broad categories, i.e. study-related activities, social activities and leisure time entertainment (see Table 7.2). Most of the activities have been mentioned in the learning experience section as situations of learning or growth.

It should be noted that these reports may not be about themselves, and can be their perceptions about their peers. So the list of who has mentioned what activities is only to show there are some agreements, and does not necessarily indicates the proportion of students that have conducted a specific activity.

Table 7.2 Student activities

Category	Specific activities	Mentioned by
Study-related activities	Study-related projects	B2C, D2a, D4d
	Practice and internship	B2C, B2a, D3a
	Courses, exams	B2C, D2
	Contests	B2C, D4d, D4a
	Preparing for postgraduate entrance exam	B2
	Attending interesting lectures	B3C, B4
	Study extra contents	D2B, D2a
Social activities	Activities in student associations	B2C, B4, C1, D2B, D2a, D2B, D3a, D4d
	Volunteering	B2C, D2c
	Romantic affair	B2C, D3b, D3a, D3b
	Part-time job	B3B, C1a
	Travel with friends	C2
Leisure time entertainment Category	Sports	C3a, D2a, D3a, D3b, D3d
	Computer games (mainly by boys)	B2a, C2, D3b
	TV series (mainly by girls)	B2C, B2a, C2, D1D
	Watching movies	C3C, C3a, D1D, D4C
	Shopping (offline and online) (mostly by girls)	B2C, B2b, C3C
	Surfing online	C3C, D1, D4b
	Reading for leisure	B3C, B3B, C2, C3a, D4d, D4b

7.1.3.3 Opinions about good students

In students' opinions, a good student has certain attributes during the study and achieves good outcomes at graduation.

- *During the study*

During the study, good students should engage in study and be responsible for themselves (B1, C1B). They should have clear goals and plans, and strong willpower to implement them (B1B, D2B, D2c). In particular, they should know the requirements of future work and actively prepare for it (B1, D2):

It can't be wrong to study according to the employers' requirements when you don't know what to study. B1

Then they should actively grasp all kinds of opportunities to develop themselves, to try out and experience, to participate in projects (B2b, C2A, D1C), and also to get certain qualifications or certificates to demonstrate their skills such as English, computers, or internship experience (B2b) which are important capital in job-hunting. Some students also think that good students should collaborate with teachers (D2a), be good at their study and master the subject knowledge (B1D, B1B), and good scores in exams can be one indicator (D2C), at least no failure (B2b). In addition, good students are also expected to be good in morality and at teamwork (B1).

- *At the end of the study*

At the end of their undergraduate study, as graduates, good students should have mastered the subject knowledge and skills (C1a, C2D, C2C, C3B, D2B) which are necessary for them to be competent at work (B4c, C2A, C2D, C3B, C3a). As an external symbol, they should have got an acknowledged good job or successfully applied for further study at a prestigious institution (B1, C3a). They should also be good at interpersonal communication and dealing with basic social occasions (B4b, C1a), especially occasions related to their professional ethics (C2D). In addition, a few students also mentioned that they should have acquired a new understanding of themselves and the world, and be self-confident, not follow others blindly (C1a, C2A). They should be able to combine school and society (C2c).

7.1.3.4 Summary of students' self-perception

The main concerns of the students consist of both job prospects and opportunities for further study; however, further study is often considered as a means to achieve better job prospects. They have been conducting a variety of activities among study-related activities, social activities and entertainments. In their opinion, good students are those who have both made full use of opportunities to develop themselves during study and started a prosperous career at the end of the study. These reports on self-perception by the students show that they are consciously concerned most about their job prospects, while they also cherish all the opportunities to develop themselves.

7.1.4   Opinions about exams

At Chinese universities, examinations usually take place at the end of the semester, and most of them are in the form of paper-and-pencil tests.

7.1.4.1 Difficulty

Most students think that exams are not difficult, and that they don't have to study hard in order to pass (B2, B3B, B4a, C3, D4a, D4C). Almost all the student participants agreed that the last study period of the semester (about one week to one month) usually determines the exam result (B2b, B3a, B3B, B4, C1, C2, C3, D1, D2). With a short intense study period before the exam, the students can get a good score, and at least have no problem passing. This is because usually the teachers will limit the scope to be tested in exams (B2, B4, C1), and the students only have to pay attention to a small part of the studied contents. Besides, the exams are mainly to do with memorization (B2C, B3C, B4b, C2), and sometimes the items tested are more or less the same year after year (D4C). With regard to not studying very well at the usual time and trying to make a concentrated effort just before exams, D1b said,

Getting a high mark is not so probable, but there would be no problem passing. D1b

7.1.4.2 Validity

Here, validity refers to whether and to what extent exams reflect the level of study achieved by students and the results they get in exams. In the students' opinions, exam results have a weak correlation with study (B3a, B4a, C1B, C2, C3). There are students who are not so good at studying but get good results, and students who have studied well all the time and yet obtained less than good results. Although students who are good at studying may not get the highest scores in exams, they can't be very bad. There are some special techniques for exams such as guessing/detecting the key points to be tested (B2, C2, C3), and paying attention only to the key points given by the teacher (B3a, B4). Such techniques can also be transferred to future exams (C1). Some engineering students also reflected that exams cannot test some of the engineering skills needed to produce something (D3a, D4b):

Exam results reflect students' ability to deal with exams, or make a concentrated effort, not necessarily the level of their study. Students who are really good at studies may not get the top places in exams; but they couldn't get very bad results. C1

7.1.4.3 Impact

Since most students believe that the last period (one week to one month) of the semester determines the exam result to a large extent, exams tend to impact their study in this period (B2b, B3a, B3B, B4, C1, C2, C3, D1, D2).

Since consequences of failure in exams are serious, e.g. to be put into a lower grade or failing to get the diploma, exams eventually do push the students to study (C2C, D1b). In this case, the students would just pursue a minimum pass score (B4c). And for some students exams are the only reason for study (C3C, D2B). But the effect can be very limited, a point made in the Group B1 discussion:

If you study something only for exams, you will soon forget it afterwards; but if you study out of your own interest/will, you can recall it in future. B1

In extreme cases, even the pressure of exams does not work at all, as claimed by D3b:

In order not to waste your time on some useless courses, it is necessary to cheat in exams sometimes. D3b

The positive impacts of exams are also reported by the students. Exams push students to learn (B4a, C2, C3B, C3C, D2B), and they provide students with an opportunity to review and systemize what was studied over the whole semester (B3C, B3D):

The only function of exams is that they give us a small warning – don't be too lazy, there is still the exam to pass. B4a

Some students think that it's not fair for students studying all the time to get the same or even a lower score than those employing a short intense study period. So some of them might be motivated not to study so hard all the time (B2a, B4), and study only the exam-related contents. This provides students with a mechanism for escaping from study: not really studying but getting a good exam result as a symbol of good study.

After exams the students are usually just given only a score, and they get little feedback on their study from exams (B2, B4, C3, D4). Normally, they would not go back to review the contents where they haven't studied very well.

7.1.4.4 Summary of students' ideas about examinations

University examinations do not seem to be very difficult for the students, and their results have a moderate correlation with the level of study achieved by the students. But they still serve as important activators for the students to study.

7.2 Teachers and teaching

When discussing good teachers and teaching, the students greatly emphasized teachers' caring attitude toward teaching and being responsible for the students (B2, B3, B4, C1, C2, D1, D3, D4). In addition, the teachers' mastery over the subject knowledge (B2, B3, B4, D1, D4), sharing their own experiences (B4b, C1E, C1D, C1B, C1C, C2 D3a), employing proper teaching tactics (B3, C3, D1, D2), being strict with students (D4), etc. are also characteristics of good teachers and good teaching.

7.2.1 Care about teaching and being responsible for students

In the students' opinions, a good teacher, first and foremost, cares about teaching and is responsible for the students. The teachers who care about teaching prioritize teaching on their agenda and take teaching seriously in a non-casual way (B2, B3, B4, C1, C2, D1, D3, D4). They constantly update the teaching contents and adjust their teaching according to specific situations and students; they pay careful attention to students' learning activities and give pertinent feedbacks. They are patient in guiding students and sensitive to the differences between students:

Interviewer: What changes do you think can be made by your teachers that will promote your learning and personal growth?
B2a: The teachers should take teaching seriously. When the teacher cares about the course and takes it seriously, so will we. He/she will take some steps when he/she cares about the course. It's quite obvious.
B2b: Yes.
B2C: Yes, very obvious.
B2a: He/she may be stricter, or set more assignments, etc. And we can feel it.
B2C: Not necessarily a lot of assignments. But the assignments must be very frequent and with timely feedback. He/she is probably in the habit of

arranging some time slots for each student in the class to have an individual or two-students-together talk with him/her each semester to summarize the student's study so far or to guide the student on how to proceed in the following study.
...

B2C: *When a teacher loves this profession, his/her passion will also influence us to love it. Some teachers give us the impression that teaching is only a task/burden for him/her. And this makes us feel that studying the course is a task/burden for us.*
Interviewer: *Then what's the difference between the teachers who see teaching as a task/burden and those who love it?*
B2a: *We can feel it from the way he/she talks in class, his/her attitude and how he/she has prepared for the course.*
B2C, B2b: *Yes, yes.*
B2a: *Some teachers have prepared the course once and for all, never changed anything.*
B2b: *He/she teaches the course the same way all the time.*
B2C: *And makes it feel boring and tedious.*
B2a: *And sometimes we fall asleep. Some other teachers who care about the course will try different approaches to mobilize us, unlike those who see teaching as a task to complete.*

In fact some teachers are utilitarian, and put most of their efforts on projects and/or publications that bring them money and promotion, but don't very much care on teaching or academic issues (B3, C2, D4b). Many teachers are busy with publication and promotion issues and treat teaching as a perfunctory matter. They are passive in their teaching – unwilling to supervise students in internship, not teaching students anything if they are not asked by them (the students) to do so. Irresponsible teachers spend more time and energy on their own issues and teach in a casual way in terms of badly-organized teaching (unclear logic). In contrast, good teachers ask students to participate as much as possible and remind them of issues they should pay special attention to.

In addition, responsible teachers are not constrained by students' grading of teachers in terms of lowering academic standards to please them (D1, D4). They can still be strict about academic criteria. They really put their heart into the students, are sensitive to students' reactions and really try their best to help them, not just standing there reading the textbooks. The students can clearly feel whether or not a teacher has put their heart into it.

B4b thought his supervisor was a responsible teacher:

He was, at the same time, supervising students from different grades, including first-year, second-year, fourth-year and graduate students. When he had supervising meetings with them, he always asked them about their progress in life and study, then gave pertinent advice to each individual student. He shared his own experience as a student and suggested possible options to choose from in order for us to avoid unnecessary detours. B4b

Two student groups from University D, D1 and D2, mentioned a very good teacher in the following way:

He asked us to be punctual, and explained that this was very important for our future social life by giving examples of what could be the consequence of being late for work. D1, D2

In their opinion, the responsible teacher is concerned about the students' future and teaches them how to behave in society. Sometimes the students need someone to push them in order to unearth some dormant attributes in them.

7.2.2 Try to understand students and relate to them as people

Good teachers in the students' eyes try to understand students as people, and put their teaching in the context of students' lives.

First, they share with students their own experiences of learning and practicing the profession in order to show the students some practical tactics (B4b), to convey to the students precise attitudes and principles in doing experiments (C1E), to make the students aware that they have much more to learn (C1D), to make abstract knowledge more visible and reachable (C1B), not just repeat what the textbook says, which can be acquired just by reading (C1C), to connect knowledge with clinical cases, and to reorganize and paraphrase textbook knowledge with personal reflections (C2). This is also a symbol of teachers' mastering of knowledge (D3a). They teach not only through the class lectures, but also through their own behavior as practitioners of the corresponding profession, and sometimes as people (D3b). One of the University B students, who is in a teacher training program, said that:

We have unobtrusively and gradually imitated our teachers' ways of being a teacher in teaching attitude and methods, etc. B3C

Second, they try to understand students as individual people, and connect what they are teaching to the rest of students' lives, or borrow the students' own words, to understand the students by putting on the students' shoes (B3D). Besides study, good teachers are also concerned about students' lives. They know what the students are doing besides study (C2A). B4a provided an example of teachers who had a good approach taking students as people. He said:

Take our C language teacher, Mr. Yang, as an example. He held a name list of the students with pictures of them at the beginning of the course, through which he was trying to remember all the students. Three weeks later, he didn't need that list anymore, because he had already got to know all the students in class. B4a

Some students also expected the teachers to be both teachers and friends, at the same time being strict over academic issues and doing sports with them (B4a, B4b).

7.2.3 Understand the art of teaching

The students also expect their teachers to master the art of teaching. In their opinions, good teachers are good at motivating students and steering the class dynamics (B3, C1a, C2A). They may organize course activities using a variety of methods such as PBL (problem-based learning), demonstrating the speculation process, putting theories into activities, more class interaction, etc.

Students from University C reported that good teachers may use innovative teaching methods such as PBL (C1E, C1a, C3). In their PBL teaching, the teachers give a patient/disease case and related questions to promote students to look up relevant knowledge and data, and then guide students in this process. By employing this approach the teachers give students opportunities to present their findings and organize discussions, to think about the problems, to exercise. In the students' opinion they improve their ability in thinking and searching, not only just taking in what the teacher conveys. These cases make the students see the relevance of knowledge and thus motivate them to study actively. In the traditional lecture-based classes the students hoped for more class interaction to avoid the situation where classes were more like the teacher's solo performance (C1a).

The good teacher must know how to guide students in learning (D3), how to deliver knowledge to students, and how to guide students to acquire knowledge. They pay attention to the feedback from students (e.g. by asking whether they understand or not), then teach based on the level acquired by students (D2). Two groups of students from University D (D1, D2) praised a teacher who insisted on using chalk and a blackboard (instead of PowerPoint) to present the speculation process. They said that:

In this way the teacher conveyed a clear thinking process, well-organized knowledge, and made us think with him, know where to go and why to do it this or that way. This gives us no chance of being distracted. PowerPoint is not suitable for presenting this process because the click is too quick to leave time for us to go through the speculation process in our brain, and it makes it very difficult for us to follow the teacher's thinking process. D1, D2

Another tactic used by good teachers is to give an overview and schedule of the contents to be studied at the beginning of the semester, so that the students can have clear expectations and be prepared about what is to be learned (D2). Yet another tactic is to teach students the basic theories and methods first, connect theories with application to make it understandable for the students (D2B), and then ask them to apply this in their own activities (B3C, B3D).

The art of teaching, in the students' eyes, also involves being strict with students. Students do not think a teacher is good because he/she gives them an easy pass. A good teacher is strict about academic criteria (D1a, D4).

7.2.4 Mastering of knowledge

The teachers' mastering of knowledge is also highly-valued by the students. Subject knowledge is the basic necessity, and social experience (how to deal with specific social or professional occasions) is also desired. A good teacher should master the subject knowledge (B3B, B3D, D1, D4), and maybe also do good research in his/her area (B2b). In his/her teaching, he/she should not only deliver the classic theories in the subject (which is the most basic requirement) but also keep learning about the recent technological applications in companies and relate teaching to them, learning about the most recent developments in this profession and show them to students (D4b). In teaching, good teachers should make sure the basic subject knowledge is delivered to the students, and then try to share their social experiences (B4b, C1B).

7.2.5 Preferred personal characteristics

When asked about their opinions on good teachers, the students also showed preferences for some attractive personal characteristics of teachers, such as being charismatic (which can be quite different among teachers) (B2b, B3D), knowledgeable (B2b, D2c, D3), passionate (B3D, D4C) and humorous (B3D, D3). When a teacher is able to attract students and make them interested in learning the course, he/she doesn't need to resort to external pressure (e.g. calling the roll) to get students to attend classes; the students will join the classes voluntarily and there will almost always be full attendance (D1C, D4a). These personal characteristics are probably related to the lecture-based style of teaching.

7.2.6 Summary of students' opinions about good teachers and teaching

From their opinions about good teaching and teachers, we can also see a clear social dimension to the students' learning. They would like their teachers to be actively engaged in teaching, care about them and relate to them as people, and learn how to behave in their respective professions in particular, and also how to behave in society in general. In addition, the students also indicated their preference for active participation in teaching and learning activities, and real-life relevance of the study.

7.3 Institutions and higher education

When asked about their opinions of a good university or the quality of higher education, the students talked mostly about teachers and the guidance they acquire (B1, B2, B3, B4, C1, D2, D3, D4), the learning atmosphere (B1, B3, B4, C1, C3, D1, D2), job prospects (B1, B4, C1, C3, D3), opportunities for all kinds of academic and social activities (B3, B4, C1, C2, D3), regulations and managerial culture (B2, B3, B4, D3, D4), infrastructure and facilities for study, sports and living (B1, B2, B4, C1). To a large extent, these points about a good university made by the students are based on their experienced dissatisfaction with the actual situation.

7.3.1 Teachers and guidance

In the students' opinions, a good university must have good teachers (and mentors), and offer them guidance in both academic and non-academic issues. These teachers are responsible for students (in the current situation, some teachers consider teaching only as a burden and finish it in a rush) (D3a), put their focus on teaching (charges teachers with less other tasks), and teach the students not only with words but also through their own behaviors (e.g. the tactics for dealing with conflicts with patients) (C1). The university arrangements should enable and support frequent student–teacher interaction (for example, lower student–staff ratio, the teachers living on campus) (B1, B2, C1a, D2). Besides academic issues, the students also expect to receive guidance in professional ethics (B3, B4, C1, D3), practical tactics, possible career tracks, etc. Some of the guidance should be specific to individual students and help them develop in accordance with their dispositions (the current situation is that some students feel like they are the standardized products of a factory assembly line) (B2, D3)

The students reported that almost all the teachers gave them the impression that they were busy all the time – they have all kinds of things to do and are seldom in their office; and it's embarrassing to interrupt/disturb the teachers by raising questions with them after class (B1). At the same time, they expect guidance from the teachers:

We have a very limited horizon and see only a small world. And thus we are probably not aware of the potential directions for our development, the books that need to be read, or things that need to be done. We are too young and need someone to guide us. B1B

Another student from the same group gave an example of a teacher who was responsible in terms of guiding students:

Some of the responsible supervisors have meetings with the students they supervise once every two weeks, and ask them about the changes in their lives and study, then give them advice on what they need to do or which books need to be read, or guidance like this. ... I might have done nothing, but now since my supervisor often asks these questions, I have to do something. B1D

Since most of the students are not familiar with the programs before they start their university study, they also expect program-specific guidance in career development in the form of a combination between program introduction and potential career tracks (B2, D3).

7.3.2 Student culture and learning atmosphere

A good learning atmosphere is also regarded by the students as an important aspect of a good university. By this they mean mainly that the student sub-culture advocates study, values study very much (D2), and has a positive attitude towards life (C1a, C1B); and students are active in study, clear about what to do (C1E, D3) and can be seen studying everywhere across the campus (D1b). The learning atmosphere also has something to do with teachers' engagement and priority in guiding students in learning (D2).

A female student from University B explained the influence of the atmosphere:

If you are in a dormitory or class with a good learning atmosphere, you will achieve a different level collectively, because when everybody around is studying, it's embarrassing not to study; when everybody is studying, it would be embarrassing if you are sleeping or surfing online; you are bound to study together with them. And this is good for your personal development. B1B

In a good learning atmosphere, the students not only value study very much and study hard themselves, they are also glad to help each other with study-related issues, and in particular, senior students are willing to help the juniors with study challenge and are glad to share their own experience (B1, B2, D1).

The students also reported some examples of a bad learning atmosphere. One male student from University B gave an example of his own experience of being laughed at when he went to self-study:

I remember that when I went to the classroom from my dormitory at about eight o'clock one morning of last year, I encountered a senior student, and he said to me, 'You are too serious [about study].' I thought it was he who was too unserious; it's not my problem. This is really a bad learning atmosphere. B4a

There is also evidence about the learning atmosphere from the topics discussed on the student BBS. From there the concerns of the students can be gleaned. If the topics are always trivial, the students at that institution must be very bored (B3).

7.3.3 Physical conditions (infrastructure and facilities)

Good universities, in the students' eyes, should also have good physical conditions. The infrastructure and facilities should be able to provide students with a resourceful and enjoyable place to live and to study. For living, there should be a beautiful campus with trees and lawns (C1a), good canteens taking care of students' health and nutrition (B2a, C1a), good accommodation conditions (six students lived in a 20 m2 room at University B at the time of the interview) (B4c), and good sports facilities. For studying, there should be classrooms equipped with modern educational technologies such as multimedia facilities (B4a), a voluminous library (C1a), and enough space for self-study (B1, B2, B4). The management of these facilities and infrastructure should be targeted at the convenience of the students, not so much at the low cost or convenience for administration.

When higher education expansion brings many more students into each university, the facilities and resources should be increased accordingly (the current situation is that there are too many students with relatively insufficient resources) (D4). With regard to accommodation conditions, C1a expressed his wish for fewer people living in one dormitory room and having more private space for study and living.

7.3.4 Rules, regulations and managerial culture

The administration is also an important aspect in the students' views of a good university. With regard to the students, a good university should have strict academic standards and encourage the students to study hard, and at the same time provide them with easier opportunities to change programs (since most students are not familiar with the programs before they start university study) (B3), and offer them all kinds of courses from which they can freely choose (the current course system has put too many restrictions on students in terms of course selection) (B3, D3). The expansion of higher education and the almost guaranteed diploma make the students feel little pressure on their studies and some of them even get degenerated because they know they are unlikely to fail in exams and thesis assessment (D3). So a good university should have strict academic standards for both enrolment and graduate certification, or at least be strict about graduate certification (B4, D3):

There are students who have got their diploma without even knowing the basic subject knowledge. D3a

With regard to teachers, a good university should direct the teachers' energy into teaching, e.g. change the current teacher evaluation system to a system with more weight on teaching and less weight on numbers of publications (the current system concerning bonuses and promotion has put too much weight on research in terms of research projects and publications) (B3). There should be some kind of censorship in the conduct of examinations, for example by letting other teachers of the same discipline design the examinations (the current examinations are both designed and marked by the same teacher who teaches the course). Then the evaluation of teachers can be based on the results of student assessment. The professors should be the main part in the governance of the university.

In addition, the managerial culture at a good university should center on the students; its arrangements should prioritize students and their study. The participants of Groups B2 and D4 gave several examples of a managerial culture which is not centered on the students and their study:

Whenever there are any events [that need to use classrooms], the classrooms for self-study can be occupied, regardless of the needs of the students. Preparing for the postgraduate entrance examination is very important for both the students and the institution, but we don't have special rooms for this. ... The way our university is managed doesn't tell us that study is an important and prioritized issue which cannot be casually disturbed. B2C

When we moved to this new campus which received quite a lot of financial support, we acquired facilities such as better laboratories. But my impression is that the laboratories only open twice a year. We students haven't enjoyed much from this financial support. The facilities are like a kind of decoration. Some laboratory administrators told us that 'The equipment is easily depreciated when used too much. And the cost for maintenance is quite high.' The equipment is there for the students to use, that is its raison d'être. D4a

B2a summarized the current managerial culture as not according to the students' needs but for administrative convenience. In addition, at a good university the students' voices should be heard in some decision-making processes (D4).

7.3.5 Career prospects for graduates and reputation

Where the graduates go is also an important indicator of the quality of a university. Graduates being employed by employers who treat them well (B4b, C3a) or being accepted for further study by prestigious institutions, especially foreign institutions, are symbols of a good university. The employment rate is also of great concern to the students. In addition, some students are also concerned about whether the job they have got corresponds to what they studied at university, and whether they are competent in the job (C3).

A good reputation is also important for a university to be good. As it is often difficult for people outside the academy to assess the academic level of a university, they (including employers and parents) tend to evaluate a university by its reputation and the university rankings (B1, D1D, D3). A diploma from a well-reputed university is often a necessary condition to stand out from the pile of CVs at the recruiters' desks. According to the discussion in Group D3:

Often the recruiters only look at the title of the university to do the first screen since there are a lot of graduates applying for jobs. D3

In the discussion of Groups C1 and C3, some students also mentioned the contribution made to society or the profession by students graduating from a certain university as an indicator of its quality (C1C). But other members of the group pointed out the difficulty of assessing this societal contribution – it couldn't be measured by the degrees conferred (numbers of graduates) or positions held by its alumni; that's only a shadow and not substantial (C1a).

7.3.6 Opportunities for all kinds of activities

Many students also view a good university as a place which provides them with plenty of opportunities to experience all kinds of activities, to try out different things and find out what they want, to practice and enhance their competences, to widen their horizon. These opportunities include participation in all kinds of student organizations such as the student union, associations, clubs, troupes, etc. (B4b, C2, D3), internships (B2b, C2, D3), recreation and sports events (B2, D3), and international or inter-institutional exchanges (C1C, C2).

7.3.7 Summary of students' opinions about a good university

Students' opinions about a good university also reflect their expectations on university study. From their reports, we can see that the students expect university education not only to offer them the subject study, but also to bring them better job prospects and abundant learning opportunities to facilitate their development and growth as a person in society.

7.4 Summary of quality from the student point of view

From the above report of findings from student focus groups on students and learning, teachers and teaching, university and higher education, it can be seen that students have the following main concerns over the quality of higher education:

- **Their own personal and career development, employability**

The students' concerns over employability can be seen from their consideration of the job prospects of the study as the most important factor in university and program selection, their keeping an eye on the job market and adjusting their study accordingly, their opinion of regarding getting a good job as one important aspect of good graduates, and their consideration of the job and career prospects of graduates and reputation as one aspect of good universities.

The concern over personal and career development can be mainly seen from their learning experience (knowledge and skill mastering, character and identity formation), consideration of faculty quality and learning atmosphere in university and program selection, opinions about good students (those who are good at study, have mastered knowledge and skills, are competent at work, have a new understanding of themselves), opinions about good teachers and teaching (being responsible to the student and competent at teaching), and opinions about good universities (those with abundant learning opportunities – with responsible and competent teachers guiding them in both academic and personal development issues, a good learning atmosphere, good infrastructure for learning, and supportive management).

- **Teaching staff's commitment to teaching and students, teaching competence, knowledge**
  Students' concerns over the teaching staff's commitment, competence and knowledge is directly shown in their opinions about good teachers and teaching, their learning experience in classes, and their opinions about a good university in terms of the teaching and guidance offered. A few students also showed this concern in their university and program selection.

- **The institutional arrangements in terms of learning opportunities, learning atmosphere, infrastructure, managerial support, and institutional reputation**
  The institutional arrangements such as learning opportunities and atmosphere, infrastructure and supportive management are reflected directly in their opinions about good universities, and also in their university and program selection, and learning experience (opportunity to do projects, internships and application, etc.). The institutional reputation concerns the students mainly in their university and program selection, and ideas about good universities.

In conscious choices, the students are more concerned about probable consequences such as job and career prospects; while they are in the process and actually experiencing it, they are more concerned about the quality/characteristics of the experience, mainly the dimensions of the sociality, real-life relevance, and activeness, which help to maximize their development. We can see that the two main concerns of the students are from different sources of requirement, i.e. one external, the job market and employer, and the other from the students themselves to develop and unfold their personalities. Therefore there could be, probably, some kind of inconsistency or even tensions between them, which I will look into further in the next chapter.

8

Exploring the tensions in quality of higher education

In Chapter 3, I outlined a three-perspective (i.e. policy, organization, education) framework to approach the issue of quality in higher education, and pointed out that the educational and organizational perspectives are under-researched in the current quality-related studies. Based on my empirical findings, this chapter tries to initiate a discussion on the quality of higher education from the educational and organizational perspectives. It begins with a summary of the research findings presented in Chapters 5, 6 and 7. This is followed by a discussion around the points of view of the institution, the teaching staff, and the students respectively. Finally, the discussion will bring the three points of view together to indicate the complexity of the quality issue in higher education.

8.1 Summary of perceptions of quality of higher education

In Chapters 5, 6 and 7, I presented respectively the findings from content analysis of institutional self-evaluation reports, teaching staff interviews and student focus groups. The questions addressed to them are centered on their perceptions of quality in higher education in terms of higher education institutions, teaching staff and students. The main findings of their perceptions in relation to these three elements are summarized in the following figure (see Table 8.1).

From the figure, it can be seen that quality of higher education, from the point of view of the institutional self-evaluation report, mainly focuses on the externally visible parts (or parts that can be easily made visible) of all the three actors, i.e. the institutional operational documents, qualities of infrastructure and facilities, awards, reputation, key research projects held by staff, student employment rate, etc.

From the teaching staff point of view, quality of higher education means both the quality of academic work and its administrative recognition such as promotion. In their academic work, they highlight both their own teaching and research competence. They regard institutional support for teaching and research as the main reference for institutional quality, and student development as student-related quality.

Table 8.1 Quality of higher education: perceptions of the institution, teaching staff and students

Quality of Source/actor	Higher education institution	Teaching staff	Students
Institutional self-evaluation report	*Visible comprehensiveness of operational documents, awards, reputation, quantities of infrastructure and facilities*	*Awards, publications, reputation, research projects*	*Awards, employment rate*
Teaching staff	*Infrastructure, policy and managerial support for teaching and research*	*Teaching competence, research, promotion*	*Personal and professional development*
Student	*Learning opportunities, learning atmosphere, infrastructure and facilities, managerial support, reputation*	*Responsibility for teaching and students, teaching competence, knowledge*	*Personal and career development, employability*

The students mainly refer quality of higher education to their own development and expanding life chances through employment. Regarding their development, they focus on the learning opportunities offered by the institution, and the teaching staff's commitment to and competence on teaching. In relation to employment they also focus on the job prospects after studying a certain program at a certain institution.

However, in their own activities there are different orientations to pursue quality for each of the three actors. There are often tensions between these orientations.

8.2 The orientations and tensions in the pursuit of quality

In the analysis of my empirical findings, there emerged two main distinguishable orientations in their pursuit of quality of higher education for each of the three actors that I have investigated – the institution, teaching staff and students. There are some tensions between those orientations for each of them respectively; and these tensions form their main concern in the pursuit of quality. For higher education institutions, it is between educational prosperity and organizational prosperity; for the teaching staff, it is between academic prosperity and administrative prosperity; for the students, it is between development and employment. However, these different orientations in tension are not in contradiction with each other – they compete with each other for priority and attention but are not radically incompatible with each other.

8.2.1 Higher education institutions – educational or organizational prosperity?

As articulated in Chapter 5, the tension facing higher education institutions in quality assurance is between educational and organizational prosperities.

8.2.1.1 Educational prosperity as quality

In the pursuit of educational prosperity, higher education institutions are devoted to the educational enterprise, which focuses on the development of students, and in association with it, the intellectual development of certain disciplines. It mainly involves an educational process that advances the student to levels of reasoning that enable him or her to critically reflect on experience, as Barnett (1990, p. 202) has argued. In this pursuit, the institutions constantly examine their educational arrangements and the impacts they have on the students (Barnett, 1992, pp. 6–7). Or according to Biggs's (2001) more elaborated framework, the institution should constantly reflect on: (1) its quality model, which makes explicit its espoused theory of learning and teaching based on the scholarship and knowledge base of teaching. It should be the basis on which teaching-related decisions are made; (2) its quality enhancement practice, which is continually reviewing and improving its current practice in response to new content knowledge, changing student populations and changing conditions in the institution and society; (3) its quality feasibility, which tries to identify the impeding factors for quality teaching, and tries to remove the impediments.

8.2.1.2 Organizational prosperity as quality

As demonstrated by the content analysis of institutional self-evaluation reports, in the pursuit of organizational prosperity, higher education institutions are devoted to being socially visible and recognized as 'good' in order to gain the resources and legitimacy for their survival and prosperity. Like other organizations in society, higher education institutions are also struggling for survival by raising resources from the environment. An effective organization is one which responds to the demands of its environment according to its dependence on the various components of the environment (Pfeffer & Salancik, 2003, p. 84).

The content analysis of the institutional self-evaluation report showed that the institutions had strategically highlighted their visible parts such as the comprehensiveness of their internal operation documents and awards, in order to construct a good-looking image. This is more to gain social recognition, and thereby seek legitimacy to raise resources, than to address the actual educational activities.

8.2.1.3 The tension between educational and organizational prosperity

It seems that higher education institutions as organizations are facing a situation similar to the individual dilemma between expression and action articulated by Goffman (1959, p. 33) – 'those who have the time and talent to perform a task well may not, because of this, have the time and talent to make it apparent that they are performing well.' His quotes from Sartre (1956, p. 60) more vividly illustrated this dilemma: 'The attentive pupil who wishes to be attentive, his eyes riveted on the teacher, his ears open wide, so exhausts himself in playing the attentive role that he ends up by no longer hearing anything' (Goffman, 1959, p. 33).

With the accountability requirement, institutions have to show to external stakeholders, especially the main sponsors such as the government, that they are doing well (at least it looks so) and the resources offered to them are efficiently used and producing good results, or at least not wasted. Therefore, they can have the legitimacy to get further access to resources. But this show and demonstration will probably distract the institutional attention and energy from actually doing the educational work, especially in a situation where the external stakeholders don't (or are too busy to) fully understand the nature of the work, and where the quality of the outcome is difficult to directly demonstrate. For example, during the Chinese institutional evaluation of undergraduate teaching, many universities established specialized offices to coordinate the evaluation, which did nothing educational; and a large amount of energy and resources was spent on evaluation (Ji, 2009; Lee, Huang & Zhong, 2012). The institutions in such a situation resort to the already legitimized ways of showing quality such as awards to show the quality of their work. Or they resort to showing the conditions which are (or they think they are) commonly believed to be associated with quality results, for example, comprehensive and clear internal operation documents, and large quantities of resources such as buildings, laboratories, libraries, famous professors, research funds, etc.

In extreme cases, these distractions may lead to goal displacement (Merton, 1957, pp. 199–202) where the means to achieve the educational goals become ends in themselves. For example, to raise more resources can become an end in itself because sometimes resources are a symbol of quality, as is the case in some university rankings[15] and also in Chinese undergraduate teaching evaluation where resources are an important indicator. The institutions can be encouraged to produce more and more documents on internal operations without any significant improvement in educational activities. In addition to that, whether it will lead to educational improvement by pinning everything down with documents or detailed procedure is in itself questionable, since educational activities involve lively human interaction and may need flexibility.[16] In addition, different types of awards, as already analyzed in Chapter 5, have their own focuses and are not systematically consistent with the educational enterprise; and usually the awards only involve a very small proportion of staff or students, and thus cannot be used as the overall level of quality.

The institutions may often find it difficult to show their efficiency or the 'goodness' of their activities. The grassroots teaching and research activities are usually invisible to the government officials. The educational impacts on the students may not be immediately obvious, the unemployment of graduates may be caused by the harsh economic conditions, and there may be many unsuccessful research experiments. Thus, higher education institutions as educational organizations have to pay attention to both doing the educational job and demonstrating to the external world that they are doing it well. Sometimes, these two kinds of work may compete with each other and cause tension for the institutions.

The tension between the educational- and organizational prosperity lies in the fact that it is often difficult to show the educational impacts directly to the external world, especially the resource providers. Trying to be good-looking inevitably distracts from the energy that can be used to improve

---

[15] For example the Net Big Ranking, the 'Chinese University Evaluation' ranking by Wu Shulian in China, and U.S. News Best Colleges rankings.
[16] Here I am not indicating there should not be any regulations at all. Some level of written regulations is often necessary, for example, those regulations that give some basis for the rights of the weak parts (most often the students) when confronting the strong parts (most often the institutions).

and actually be good, because some of the energy has to be used in finding out what good-looking looks like and pretending to be so.

8.2.2 Teaching staff – academic or administrative prosperity?

In pursuing quality of higher education in their own work the teaching staff are also facing a tension between two orientations, i.e. the administrative prosperity and academic prosperity.

8.2.2.1 Academic prosperity as quality

To pursue academic prosperity is to devote oneself to the development (both personal and professional) of the students and the development of certain intellectual areas. This concerns the knowledge and skill level, the learning ability and the growth of the students. Based on these concerns, academics commit themselves to teaching and make efforts to enhance their own teaching competences. Besides teaching, they may also be concerned about the advancement of their subject or area through research where they pay attention to intrinsic characteristics of research, i.e. the rigidity, the content, the interaction with peers, the significance of certain research in the development of their research area or subject, etc. By doing research the academics tend to deepen their understanding of the subject, and thus may also enhance their teaching.

8.2.2.2 Administrative prosperity as quality

Administrative prosperity here means performing well in the university administrative sanction/incentive system, i.e. good results in staff evaluation, promotion, bonus, awards and financial support, etc. Staff that are administratively prosperous usually get higher salaries, and gain access to more resources, probably also more freedom, to conduct their activities.

8.2.2.3 The tension between academic and administrative prosperity

The tension for the academic staff in pursuing quality is between academic- and administrative prosperity, rather than between teaching and research originates from the teaching staff interviews. In the staff's views, teaching and research are both what the academics think they should do and what they would like to do. They see teaching and research as interrelated and mutually reinforcing, and on a very few occasions these two conflict (BCL, BLM, BZX, CFC, CLN, DCW, DGJ, DXY, DYL, DZM, DZQ), especially when the courses they teach and their research interests coincide with each

8.2 The orientations and tensions in the pursuit of quality

other. Research makes them master a specific subject better and thus improves their teaching. They can bring their research contents to enrich teaching (for example, some research can be used as cases to make teaching more vivid), or bring students into research projects.

The problem, as seen from the staff's point of view, is that the university policy has been putting too much emphasis on research and too little on teaching. They are far more encouraged and rewarded by doing research publications than teaching, as reported in Chapter 6. Therefore most of the teachers put more energy and time into research. Sometimes this causes inner conflicts within the staff themselves as teachers. In the Chinese culture, teachers are highly respected in society, since they are the ones who 'transmit Dao (or Tao, Confucian morals), imparting knowledge and resolving doubts', and 'cultivate people's soul' (Hui, 2005). Academics have a strong identity as teachers. The conflict is best illustrated by the following quote from my interview:

We teachers should improve our teaching even though the current university policy does not put so much weight on teaching in the teacher evaluation and promotion system. DGJ

BZK from University B criticized the fact that some universities are becoming research institutes where 'university' only exists in their titles.

The way university policy emphasizes research is also problematic. In the current Chinese higher education administration system, it seems that the number of publications, level of the journals (indicated by SCI admission, impact factor, etc.) where articles are published, the number of national key research projects, and amount of research funding matter much more than teaching and the intrinsic quality of the research. So pursuing more publications is the most rewarding way of achieving this administrative prosperity. And publishing articles in SCI-indexed journals can be an end in itself, no matter whether the research done is of any relevance to the host society and what the intrinsic characteristics of the research (mentioned above) are.

The teaching staffs have to conduct their academic activities, and also earn their living, under the university administrative framework, which offers them the resources and other support necessary for academic activities. To get promoted and rewarded in the administrative system, i.e. administrative prosperity, and gain access to more resources and support from the system, one has to not only follow but also perform well according to the administrative rules and incentives. But if one follows the administrative incentives too much, it can lead one into academic withering. The administration may not be academically sensitive, and may misunderstand the inner logic of academic work. In the Chinese case provided here, following the administrative incentive can lead to one putting little time and energy into teaching, as already documented in Chapter 6; and with regard to research, it can lead to one publishing immature research results (even plagiarism as has already happened), focusing on short-term research and avoiding more demanding research which might lead to greater breakthroughs for the discipline, or selecting externally initiated research which is provided with more funding but may not be academically significant (e.g. Jiang, 2009; Wang, 2011).

From their reports on time-energy allocation, challenges and support expectations in teaching, and opinions about good universities, it can be clearly seen that there is tension between the teaching staff's pursuit of academic and administrative prosperity. They need more institutional recognition for their work in teaching.

8.2.3   Students – development or employment?

In pursuing quality of higher education the students are facing tension between two orientations, i.e. (personal and career) development and expanding life chances through employment.

8.2.3.1   Personal and career development as quality
In pursuit of personal and career development from higher education, students devote themselves to knowledge and skill mastering, professional competence, personal growth and identity formation (including developing meaningful relationships with people), as discussed in the learning experience section of Chapter 7. Students in this pursuit regard institutions with abundant learning opportunities as good institutions, and teaching-committed and competent staff as good staff.

8.2.3.2   Expanding life chances as quality
Current undergraduate students in China have an urgent concern over trying to expand their life chances, by increasing their employability in the job market from higher education. In this pursuit, they select a university and program according to the corresponding job prospects (employability) after the study. During the study, they pay attention to which items are listed in the job recruitment requirements, and study accordingly, e.g. to

select job-relevant courses. They also try to get extra certificates which might make them stand out in job competition. Students in this pursuit regard institutions with a good employment reputation as good institutions, and staff familiar with current application as good staff.

8.2.3.3 The tension between development and expanding life chances

Personal and career development is more about the formation or unfolding of a self, individual growth and evolving as an individual in society (Jarvis, 1992), to live a meaningful life. Life chance is more about survival in a fiercely competitive job market and improvement of one's social and economic status, which requires the individual student to sacrifice part of the self and fit into external requirements, for example, pass exams, get good records and extra certificates, develop character traits that are popular in the job market, etc.

Before the start of higher education expansion in 1999, there was only a small proportion of people who could get access to higher education. Thus, a university diploma could almost guarantee a decent job and a better life chance than the majority of the population. Since the expansion, when more and more people have been gaining access to higher education (cf. Chapter 2), the relative value of the diploma has also been decreasing. An undergraduate diploma has almost changed from a luxury or privilege to a necessity in the job market. The students' concerns over job prospects has also been increasing.

The conscious concern of most of the current Chinese undergraduate students is more about expanding life chances, or sometimes survival in a competitive society, than personal and career development. They are trying to be well-adapted to the social system and improve their social and economic status. In fierce competition for job opportunities, the ones who fit the recruitment requirements and the existing job positions tend to get the precious opportunities. But if they go too far in this direction there is the risk of losing their selves and becoming automatons, as coined by Erich Fromm (1941, pp. 185ff). In the extreme case, 'He thinks, feels, and wills what he believes he is supposed to think, feel, and will.' or 'I am nothing but what I believe I am supposed to be' (Fromm, 1941, p. 254). In this sense the students lose their human authenticity to just perform the role or several functions they are supposed to fulfill, and become exactly what is needed in the job market and thus cogs in the vast economic machine. They might be quite competent at certain specific work, but lack the self-direction and ability to choose and adjust, which is often necessary in modern work.

8.3 The dilemma between efficiency and sustainable development

The tensions faced by higher education institutions, the teaching staff and students, as illustrated above, seem to echo the dilemma between efficiency and sustainable development. Organizational prosperity pursued by institutions, administrative prosperity pursued by teaching staff, and employment pursued by students are all devoted to being well-adapted to their current environments respectively, to immediate or short-term prosperity, i.e. efficiency. But if they go too far in this efficiency direction, they are risking losing their opportunity for sustainable development, i.e. this may lead higher education institutions to goal displacement (beginning to do something other than the reason why they are established), teaching staff to academic withering in terms of both teaching and research, and students to lose their selves and become automatons.

However, the two main different orientations encountered by each of the three actors in their pursuit of quality of higher education are not, in principle, in contradiction to each other.

Quality assurance and other accountability mechanisms are established to promote higher education institutions to better complete their educational missions. Thus, organizational prosperity is ultimately pointing in the same direction as educational prosperity. The reason why there is tension between them lies in the current ways realizing institutional accountability and the lack of knowledge about the proper ways to achieve it.

The situation faced by the teaching staff is similar to that of the institutions. The institutional administration is established to support and facilitate academics to do their academic work; and the incentive system regarding awards and promotion is supposed to promote academics to better complete their work. Thus, the administrative prosperity is ultimately pointing in the same direction as the academic prosperity. Only the current administrative ways of promoting the academics are distorting to some extent. There is a lack of articulated knowledge about the nature of academic work or proper ways to evaluate and promote the academics.

For the students, personal and career development and expanding life chances through employment are actually both pointing in the direction of

8.3 The dilemma between efficiency and sustainable development

personal prosperity and living a meaningful life. Only that the social setting is there which the individual student has to adapt to some extent in order to be employed, and then they can have the opportunity to continue their development and unfolding of the self. The tension for the students lies in the increase in social competition for opportunities and the difficulty in setting the balance and/or priority between those two orientations.

For all the main higher education actors investigated in this study, i.e. the institutions, teaching staff and students, there are tensions in their pursuit of quality. However, these tensions are not a matter of fundamental contradictions but rather of distorted emphasis under certain institutional and socio-economic conditions.

9

Conclusion:
Toward a more inclusive understanding of quality in higher education

This chapter tries to wrap up the whole study reported in this book and to push further the discussion over the quality issue in higher education.

9.1 Summary of this study

In the project reported in this book I have looked into the quality issue in higher education, which is closely related to the expansion of higher education and new mechanisms of accountability employed by the government. I have constructed a framework of three perspectives for approaching quality in higher education, i.e. the policy perspective, organizational perspective and educational perspective. The policy (or government) perspective, which is mostly involved in current research, looks into quality of higher education in terms of how higher education institutions can be promoted and facilitated to realize their social roles and be held accountable for their utilization of public resources. The organizational perspective, which takes the points of view of the institutions, looks into quality of higher education in terms of how institutions as organizations can survive and prosper in relation to quality assurance. The educational perspective, based on the points of view of the teaching staff and students, looks into quality of higher education in terms of how higher education has facilitated students' development and growth. Then, I have argued that the latter two perspectives should be paid more attention to in quality-related research in order to form a more informed and inclusive understanding of quality of higher education; and this study is devoted to an exploration from these two perspectives.

The research questions I have posed are:

- How is quality of higher education perceived by the institution, teaching staff and students, respectively?
- What are the main concerns for the institution, teaching staff and students in their own pursuit of quality?

In order to address these questions, I have conducted the following empirical investigations in the Chinese context:

- Content analysis of 53 institutional self-evaluation reports, which were produced by the institution during the ministerial evaluation of undergraduate teaching;
- Individual interviews with 19 teaching staff at three universities;
- Focus group interviews with 45 students in 12 groups at the same three universities.

My main findings are as follows.

9.2 Answers to research questions and implications

Quality of higher education is perceived by ***institutions*** in their self-evaluation reports as those aspects that are externally visible, i.e.

- Comprehensiveness of institutional operational documents, institutional awards and reputation, quantities of infrastructure and facilities;
- Teaching staff's awards, publications, reputation and research projects;
- Students' awards and employment rate.

Quality of higher education is perceived by the ***teaching staff*** as:

- Institutional infrastructure, policy and managerial support for teaching and research;
- Teaching staff's teaching and research competence, and promotion status;
- Students' personal and professional development.

Quality of higher education is perceived by the *students* as:
- Institutional learning opportunity and atmosphere, infrastructure and managerial support for learning, institutional reputation;
- Teaching staff's responsibility for teaching and students, teaching competence and knowledge;
- Students' personal and career development, employability.

Based on the findings from the first research question and the interpretation in their respective contexts, I found that the institution, teaching staff and students all have their main concerns in their own pursuit of quality in the form of a tension between two orientations. To be specific,

- For *higher education institutions*, it is between the educational prosperity and organizational prosperity. In order to gain the legitimacy and resources for organizational survival and prosperity, the institutions are devoted to being socially visible and recognized as 'good', especially when they face the external requirements of evaluation and other approaches of quality assurance. This effort to demonstrate that they are doing well with the resources provided for them may actually distract their attention and energy which might be devoted to educational activities. Thus, organizational imperative of demonstration may actually impede their effort to achieve educational prosperity.
- For the *teaching staff*, it is between academic prosperity and administrative prosperity. The teaching staffs have to earn their living and conduct their academic work, which is devoted to the development of students and certain intellectual areas, with resources and other support from the institutional administrative system. The incentives and sanctions from this system may not be so well aligned with their academic pursuit; and to perform well in the administrative system may lead to academic withering.
- For the *students*, it is between development and employment. The students, on the one hand, cherish the learning opportunities for their personal development and growth; on the other hand, they are competing for limited (good) employment opportunities for expanding their life chances, which may require them to sacrifice part of their self-development and to conform to external requirements.

These tensions which concern the institution, teaching staff and students in their pursuit of quality of higher education seem to lie in the difficulty of balancing the priority between efficiency in the current situation and sustainable development. Only when they are currently efficient to a certain extent can they get the opportunity or more advantageous support for further development; but if they are too well-adapted to the current situation, they may lose the opportunity for sustainable development. However, these orientations in tension are not fundamentally contradictory to each other; the tensions between them are rather of distorted emphasis under certain institutional and socio-economic conditions. But they are still a matter for concern about what happens to higher education under the types of quality assurance systems being pursued today.

9.3 The complexity of quality in higher education

In this study, I am not intending to provide a comprehensive picture of the quality issue in higher education. The following text is meant to highlight a few aspects I have already touched on in my study and to indicate from these aspects the complexity of quality in higher education.

9.3.1 Quality of higher education as educational quality

The title of this sub-section, 'Quality of higher education as educational quality', seems to be an unnecessary repetition. Isn't quality of higher education referring to educational quality?

As discussed above, there are different orientations held by each of the three core actors in higher education. Behind those different orientations, there are actually different understandings about the nature of higher education. For the students, higher education is not only about learning and development, it is also a mechanism of differentiation, which leads them towards different future tracks (different opportunities for further study, employment and expanding life chances). For the teaching staff, it is not only for academic development and cultivating the young generation, but also a way to earn one's living and promotion. For the institutions, it is not only an enterprise for education, but also a way to gain resources for organizational operation and prosperity. Therefore, quality of higher education is not only about educational quality and prosperity; it is also about expanding life chances for the students, prosperity in the administrative system for the teaching staff, and organizational prosperity for the institution.

If we narrow our focus to the educational quality of quality in higher education and talk about quality of higher education educationally, it should be about the students' learning and development, and the facilitation of this process. Or to use my empirical findings and put it more specifically, it is about how the teaching staffs have guided the students and how the students have engaged in their studies. Here, teaching staffs guidance refers to how they have conveyed the knowledge and skills in relation to the students' life contexts, how they relate to the students as both practitioners of a certain profession and as a more mature person in general. Students' engagement in study refers to the social dimension, the real-life relevance dimension and the activeness dimension of the students' learning experience, which, to a large extent, resonate with Illeris's (2003, 2007, 2009) three dimensions of learning outlined in Chapter 3. Thus, good-quality education is where the students can make sense of it in their life contexts, where they are interested (or motivated) and participate actively, and where the students have a meaningful relation and interaction with people, either the teachers, or peers, or people their study are dealing with (e.g. the patients for the medical students).

From this perspective, the university mainly serves as a place where the students learn about the world, and learn to evolve into a knowledgeable person in the world. The quality of a university depends on factors such as the learning opportunities it has created (whether the students are facilitated to participate actively in learning), the quality of teacher–student interaction, how it facilitates teaching staffs guidance of the students, etc.

9.3.2 Quality of higher education and social economic situation

Quality of higher education is not just an issue inside higher education; it is closely related to the social economic situation.

The rising concern over the quality issue in higher education is significantly associated with the expansion of higher education. As I have already indicated in Chapter 2, the expansion of higher education in China is not for educational reasons, but mainly for economic reasons such as to provide the developing economy with educated workers (Li, Morgan, & Ding, 2008) or to increase domestic demand (Feng & Li, 2009). The rapid increase of students enrolled in higher education has brought significant challenges to the institutions in terms of insufficient resources and faculty members. For example, big classes with more than 100 students appeared, as indicated in my interviews with the teaching staff. These big classes reduced the student–teacher interaction, which is one of the most important elements in good teaching from the teaching staff point of view.

In addition, with the expansion of higher education the value of a university diploma decreased, which increased the students' concerns over employment, as discussed above. Therefore, expansion also contributed to students' tensions between personal and career development and employment in their pursuit of quality in higher education. The situation of student employment not only has something to do with their education at university, but also, probably more importantly, has something to do with the economic condition. There would be a high unemployment rate of university graduates during an economic depression, no matter what the quality of higher education is like.

9.3.3   Quality of higher education and student engagement

My empirical findings (cf. Chapter 7 about students' learning experience) also indicate that higher education must involve the active engagement of the students. Or to borrow Barnett's (2007) term, 'a will to learn', by which he refers to 'a readiness to keep going, a willingness to open oneself to new experiences, and a propensity critically to be honest with oneself and critically to interrogate oneself' (p. 7). Based on my findings, this will is closely associated with the social dimension, real-life relevance, and activeness of the learning opportunities offered by certain educational settings. According to Barnett (2007, p. 26), the will to learn is the foundation upon which all else rests in higher education; without a will in place, no serious effort can be made in learning. The students may just make a minimum effort to pass the exams as reported in Chapter 7. Or to borrow the students' own words from my interviews,

If you study something only for an exam, you will soon forget it afterwards; but if you study out of your own interest/will, you can recall it in the future. B1

Thus good-quality learning outcomes in terms of student development must have the elements of students' commitment to and engagement in learning, which to a large extent depends on the students themselves. So higher education institutions and their staff alone cannot achieve good-quality education. They can only provide opportunities and incentives that probably lead to good-quality education; it also depends on the students' making use of those opportunities to develop themselves to achieve good learning outcomes, and thus achieve the educational goal.

9.3.4 Quality assurance and organizational behavior of higher education institutions

The possible distortion effects of quality assurance have already been discussed a lot in Chapter 5 and Section 8.2.1 of last chapter. So I just give a summary here. Quality assurance and other similar mechanisms of accountability have the potential to distract the attention and energy of higher education institutions, and lead to goal displacement in extreme cases – they might lead the institutions to highlight the measurable, quantifiable and visible aspects, for example the narrow focusing on teaching staff's publications as demonstrated in my empirical findings. This may leave the other aspects unseen and with an impression of unimportant, for example, the quality of student–teacher interaction, and teaching staff's commitment to and engagement in teaching, which are cherished by the students.

9.3.5 Can quality of higher education be ensured?

Can quality of higher education be ensured? It's time for a direct look at this question.

The answer seems to be 'yes' if by quality we mean 'a smoothly operated administrative system with detailed management procedures', as indicated in the institutional self-evaluation reports by the comprehensiveness of their internal operational documents.

The answer would definitely be 'no', if we take into consideration all the discussions presented above, and if by quality we mean the human experience of learning, growth and creation. Quality of higher education not only involves the institutions and their staff, but is also closely related to the social economic situation. If we take higher education as a service, it is a special service indeed, in the sense that it involves the active engagement of its customers, i.e. the students, to produce quality results. It is always a possibility offered by the institutions, and never a guarantee or assurance. One can probably only be sure about machines, not human beings.

Quality assurance is misleading if it is taken as ensuring the results of higher education, since academic conversations that go on in higher education institutions are necessarily unpredictable (Barnett, 1992, p. 69).

If higher education is to educate the students to not only master the practical knowledge and skills needed but also the ability to critically reflect on their own experience, which makes it 'higher' education, it inevitably involves inquiry in a process of open dialogue (Barnett, 1990, pp. 202–203) and thus uncertainty. Research that goes on in higher education institutions mostly deals with explorations at the interface between the known and the unknown, which also inevitably involves many trials and failures. Therefore, it is more relevant to see 'quality assurance' as ensuring the institutional conditions for students to learn and be educated.

Then what about accountability? How can we hold higher education institutions accountable? But the why-question should probably be asked first. Why should the institutions be held accountable? And what does accountability mean?

If by accountability we mean demonstrability or the license to blame and this is out of distrust, then it is most probably distracting the attention and energy of higher education institutions from achieving their academic goals (teaching and research), as discussed above. This distrust can be self-propelling. The greater the distrust, the more urgent is the need to demonstrate, and the more energy is distracted from actually doing the work, and thus the more inefficient and untrustworthy the institutions tend to become.

But if by accountability we mean responsibility and it is aimed at promoting and facilitating higher education institutions to better accomplish their academic missions, i.e. education and research, then a 'quality *for* higher education' concept and a communicative platform would be more fruitful. Responsibility or being 'responsible', as Fromm (1957/1995, p. 22) proposed, means being able to and ready to 'respond'. So accountability in this sense refers to whether the institutions have responded on certain occasions and the appropriateness of certain responses. Thus, 'quality *for* higher education' refers to the efforts made to facilitate and enable higher education institutions to respond to the changes that are happening in society. For example, to a large extent the inefficiency of higher education institutions is associated with the various values that have been involved in their activities; thus, a good start would be to establish a communicative platform where all the stakeholders can fully communicate their needs and interests to the institutions, which offers the academics basic information for enhancing the social relevance of their activities; but they should also have the freedom not to respond to the immediate needs of society and be able to explore in different directions or further-reaching issues.

9.4 Toward a more inclusive understanding of quality in higher education

Quality assurance and accountability may also be an issue of trust (Hoecht, 2006); and trust simultaneously involves both sides, as in the question Chiang (2012) has posed: 'Is higher education not to be trusted or is the government unable to trust?' Quality as a responsibility may also require the government to dare to trust higher education institutions. Therefore, there is probably also a need to study the government to gain a more informed understanding of quality of higher education.

All organizations are probably facing the dilemma of efficiency and sustainable development (Zhou, 2003, Chapter 10). However, compared to other organizations, higher education institutions seem to serve a much wider value range including intellectual, educational, scientific, economic and cultural. They would be significantly paralyzed by being pushed to pursue efficiency in any single value. As Collini (2012, p. 198) has proposed,

Major universities are complex organisms, fostering an extraordinary variety of intellectual, scientific and cultural activity, and the significance and value of much that goes on within them cannot be restricted to a single national framework or to the present generation. They have become an important medium – perhaps the single most important institutional medium – for conserving, understanding, extending, and handing on to subsequent generations the intellectual, scientific, and artistic heritage of mankind. In thinking about the conditions necessary for their flourishing, we should not, therefore, take too short-term or too purely local a view.

Thus, with regard to quality assurance and accountability in higher education, special attention should be given to the values pursued and be careful to balance between immediate, single-value efficiency and sustainable development and a wider-range of values. Or to simply paraphrase BCG, one of my teaching staff participants, it is not incorrect to pursue efficiency and short-term goals. But there should also be space and support for those who would like to devote themselves to longer-term goals.

Appendix A: Profile of teaching staff participants

Appendix A: Profile of teaching staff

Code	University	Program	Gender	Age	Teaching year	Position
BCG	B	educational technology	male	40-50	14	professor
BCL	B	educational technology	female	30-40	3	assistant professor
BLM	B	educational technology	female	40-50	19	professor
BMN	B	educational technology	female	30-40	5	assistant professor
BZG	B	educational technology	male	30-40	11	assistant professor
BZK	B	educational technology	male	50-60	21	professor
BZX	B	educational technology	female	30-40	8	assistant professor
CFC	C	clinical medicine	female	30-40	10	associate professor.
CLA	C	clinical medicine	male	30-40	no teaching	--
CLN	C	clinical medicine	male	30-40	8	assistant professor
CMM	C	clinical medicine	male	30-40	not specified	not specified
CWB	C	clinical medicine	male	25-30	not specified	not specified
CZC	C	clinical medicine	male	30-40	14	associate professor
CZH	C	clinical medicine	female	30-40	7	associate professor
DCW	D	automation	female	40-50	more than 10	professor
DGJ	D	automation	male	30-40	at least 6	associate professor
DXY	D	automation	female	30-40	6	associate professor
DYL	D	automation	female	30-40	8	assistant professor
DZM	D	automation	male	40-50	17	professor
DZQ	D	automation	male	40-50	8	associate professor

Appendix B:
Profile of student focus group participants

Group	University	Participants	year	Major/program	Gender composition
B1	B	B1A, B1B, B1c, B1D	2nd	Educational technology	Male: B1c Female: B1A, B1B, B1D
B2	B	B2a, B2b, B2C	4th	Educational technology	Male: B2a, B2b, Female: B2C
B3	B	B3a, B3B, B3C, B3D	3rd	Educational technology	Male: B3a Female: B3B, B3C, B3D
B4	B	B4a, B4b, B4c, B4d	1st	Educational technology	All male
C1	C	C1a, C1B, C1C, C1D, C1E	3rd	Clinical medicine	Male: C1a Female: C1B, C1C, C1D, C1E
C2	C	C2A, C2B, C2C, C2D	3rd	Clinical medicine	All female
C3	C	C3a, C3B, C3C	3rd	Clinical medicine	Male: C3a Female: C3B, C3C
C4	C	C4A, C4b, C4c	3rd	Clinical medicine	Male: C4b, C4c Female: C4A
D1	D	D1a, D1b, D1C, D1D	2nd	Automation	Male: D1a, D1b Female: D1C, D1D
D2	D	D2a, D2B, D2c	2nd	Automation	Male: D2a, D2c Female: D2B
D3	D	D3a, D3b, D3c, D3d	3rd	D3a, D3b: Electronic Information Science and Technology D3c, D3d: Electrical engineering and its automation	All male
D4	D	D4a, D4b, D4C, D4d	3rd	Automation	Male: D4a, D4b, D4d Female: D4C

Bibliography

Abbott, L. (1955). *Quality and competition*. New York: Columbia University Press.

Alexander, F. K. (2000). The changing face of accountability: Monitoring and assessing institutional performance in higher education. *Journal of Higher Education, 71*(4), 411-431.

Andersen, V. N., Dahler-Larsen, P., & Pedersen, C. S. (2009). Quality assurance and evaluation in Denmark. *Journal of Education Policy, 24*(2), 135-147.

Barnett, R. (1990). *The idea of higher education*. Buckingham: SRHE & Open University Press.

Barnett, R. (1992). *Improving higher education: Total quality care*. Buckingham: SRHE and Open University Press.

Barnett, R. (2007). *A will to learn: Being a student in an age of uncertainty*. Berkshire: McGraw-Hill & Open University Press.

Basch, C. E. (1987). Focus group interview: An underutilized research technique for improving theory and practice in health education. *Health Education & Behavior, 14*(4), 411-448.

Bauman, Z. (2000). On writing: On writing sociology. *Theory, Culture & Society, 17*(1), 79-90.

Bauman, Z. (2001). *The individualized society*. Cambridge: Polity.

Bauman, Z. (2008). *The art of life*. Cambridge: Polity.

Berger, P. L., & Luckmann, T. (1967). *The social construction of reality: A treatise in the sociology of knowledge*. Harmondsworth: Penguin Books.

Biggs, J. (2001). The reflective institution: Assuring and enhancing the quality of teaching and learning. *Higher Education, 41*(3), 221-238.

Birnbaum, R. (1988). *How colleges work: The cybernetics of academic organization and leadership*. San Francisco & London: Jossey-Bass Publishers

Bornmann, L., Mittag, S., & Daniel, H. D. (2006). Quality assurance in higher education – meta-evaluation of multi-stage evaluation procedures in Germany. *Higher Education, 52*(4), 687-709.

Bourdieu, P. (1996). *The state nobility: Elite schools in the field of power*. Cambridge: Polity.

Bovens, M. (2005). Public accountability. In E. Ferlie, L. Lynne & C. Pollitt (Eds.), *The Oxford handbook of public management* (pp. 182-208). Oxford: Oxford University Press.

Brennan, J., & Shah, T. (2000). *Managing quality in higher education: An international perspective on institutional assessment and change*. Buckingham: OECD, SRHE and Open University Press.

Brubacher, J. S. (1982). *On the philosophy of higher education*. San Francisco: Jossey-Bass Publishers.

Carey, M. A., & Smith, M. W. (1994). Capturing the group effect in focus groups: A special concern in analysis. *Qualitative Health Research, 4*(1), 123-127.

Cartwright, M. J. (2007). The rhetoric and reality of 'quality' in higher education: An investigation into staff perceptions of quality in post 1992 universities. *Quality Assurance in Education, 15*(3), 287-301.
Chen, Yukun. (2009). Evaluation of undergraduate teaching at institutions of higher education in China: Problems and reform. *Chinese Education and Society, 42*(2), 63-70.
Chiang, K. H. (2012). Is higher education not to be trusted or is the government unable to trust? Analysis from a Simmelian perspective. Paper presented at *2012 Annual Conference of Society for Research into Higher Education*, December 12-14. Newport, UK.
Christensen, T., Lisheng, D., & Painter, M. (2008). Administrative reform in China's central government – how much learning from the West? *International Review of Administrative Sciences, 74*(3), 351-371.
Clark, Burton R. (1983). *The higher education system: Academic organization in cross-national perspective*. Berkeley and Los Angeles: University of California Press.
Clark, B. R. (1998). *Creating entrepreneurial universities: Organizational pathways of transformation*. Paris and Oxford: International Association of Universities and Elsevier Science.
Cohen, M. D., & March, J. G. (1986). *Leadership and ambiguity*. Boston, MA: Harvard Business School Press.
Collini, S. (2012). *What are universities for?* London: Penguin Books.
Corbin, J., & Strauss, A. (2008). *Basics of qualitative research: Techniques and procedures for developing grounded theory* (3rd ed.). Thousand Oaks, CA: Sage.
Crosby, P. B. (1979). *Quality is free: The art of making quality certain*. New York: New American.
Denzin, N. K., & Lincoln, Y. S. (2005). The discipline and practice of qualitative research. In N. K. Denzin & Y. S. Lincoln (Eds.), *Handbook of qualitative research* (3rd ed.) (pp. 1-32). Thousand Oaks, CA: Sage.
Dill, D. D. (2000). Designing academic audit: Lessons learned in Europe and Asia. *Quality in Higher Education, 6*(3), 187-207.
Dill, D. D. (2010). We can't go home again: Insights from a quarter century of experiments in external academic quality assurance. *Quality in Higher Education, 16*(2), 159-161.
El-Khawas, E. (2007). Accountability and quality assurance: New issues for academic inquiry. In J. J. F. Forest & P. G. Altbach (Eds.), *International handbook of higher education* (pp. 23-37). Dordrecht: Springer.
Ellis, R. (1993). Quality assurance for university teaching: Issues and approaches. In R. Ellis (Ed.), *Quality assurance for university teaching* (pp. 3-15). Buckingham: the Society for Research into Higher Education & Open University Press.
Feigenbaum, A. V. (1951). *Quality control: Principles, practice, and administration*. New York: McGraw-Hill.
Feng, J. & Li, W. (2009). The root of educational development lies in following educational rules [in Chinese *jiaoyu fazhan de genben zaiyu zunxun jiaoyu guilv*]. *Exploration and Free Views* (Chinese journal *tansuo yu zhengming*), (2), 53-56.
Flyvbjerg, B. (2001). *Making social science matter: Why social inquiry fails and how it can succeed again*. Cambridge: Cambridge University Press.
Fromm, E. (1941). *Escape from freedom*. New York: Holt, Rinehart and Winston.
Fromm, E. (1957/1995). *The art of loving*. UK: George Allen & Unwin, 1957; London: Thorsons, 1995.
Gan, Xingqiong & Deng, Zhen (2008). Economic analysis on the university enrollment expansion and individual educational choice [In Chinese *gaoxiao kuozhao yu geren jiaoyu*

xuanze de jingjixue fenxi]. *Educational Research* (Chinese journal *jiaoyu yanjiu*), (12), 68-72.

Gao, Yaoming, Zhang Ping, Chen Hui, Lan Lili, & Zhang Guanghui. (2009). The impact of the evaluation of the standards of undergraduate teaching work on teaching at higher education institutions: An investigative study. *Chinese Education and Society, 42*(2), 86-99.

García-Aracil, A., & Palomares-Montero, D. (2010). Examining benchmark indicator systems for the evaluation of higher education institutions. *Higher Education, 60*(2), 217-234.

Gilmore, H. L. (1974). Product conformance cost. *Quality Progress, 7*(5), 16-19.

Goffman, E. (1959). *The presentation of self in everyday life*. New York: Anchor Books/Doubleday.

Gronroos, C. (1983). *Strategic management and marketing in the service sector*. Cambridge, MA: Marketing Science Institute.

Gross, E. (1968). Universities as organizations: A research approach. *American Sociological Review, 33*(4), 518-544.

Guba, E. G., & Lincoln, Y. S. (1994). Competing paradigms in qualitative research. In N. K. Denzin & Y. S. Lincoln (Eds.), *Handbook of qualitative research* (pp. 105-117). Thousand Oaks, CA: Sage.

Gumport, P. J. (2000). Academic restructuring: Organizational change and institutional imperatives. *Higher Education, 39*(1), 67-91.

Hartley, M., & Morphew, C. C. (2008). What's being sold and to what end? A content analysis of college viewbooks. *The Journal of Higher Education, 79*(6), 671-691.

Harvey, L. (1995). Beyond TMQ. *Quality in Higher Education, 1*(2), 123-146.

Harvey, L. (2005). A history and critique of quality evaluation in the UK. *Quality Assurance in Education, 13*(4), 263-276.

Harvey, L., & Williams, J. (2010). Fifteen years of quality in higher education. *Quality in Higher Education, 16*(1), 3-36.

Heusser, R. (2006). Mutual recognition of accreditation decisions in Europe. *Quality in Higher Education, 12*(3), 253-256.

Hoecht, A. (2006). Quality assurance in UK higher education: Issues of trust, control, professional autonomy and accountability. *Higher Education, 51*(4), 541-563.

Holsti, O. R. (1969). *Content Analysis for the Social Sciences and Humanities*. Reading, MA: Addison-Wesley.

House, E. R., & McQuillan, P. J. (2005). Three perspectives on school reform. In A. Lieberman (Ed.), *The roots of educational change: International handbook of educational change* (pp.186-201). Dordrecht: Springer.

Houston, D. (2008). Rethinking quality and improvement in higher education. *Quality Assurance in Education, 16*(1), 61-79.

Hui, L. (2005). Chinese cultural schema of education: Implications for communication between Chinese students and Australian educators. *Issues in Educational Research, 15*(1), 17-36.

Idrus, N. (1996). Towards total quality management in academia. *Quality Assurance in Education, 4*(3), 34-40.

Illeris, K. (2003). Towards a contemporary and comprehensive theory of learning. *International Journal of Lifelong Education, 22*(4), 396-406.

Illeris, K. (2007). *How we learn: Learning and non-learning in school and beyond*. London: Routledge.

Illeris, K. (2009). A comprehensive understanding of human learning. In K. Illeris (Ed.), *Contemporary theories of learning: Learning theorists... in their own words* (pp.7-20). London: Routledge.

Jarvis, P. (1992) *Paradoxes of learning: On becoming an individual in society*. San Francisco, CA: Jossey-Bass.

Ji, Baocheng. (2009). There are too many evaluations at colleges and universities. *Chinese Education and Society, 42*(2), 52-55.

Jiang, Yin. (2009). Thoughts and countermeasure to deal with academic corruption and misconduct [in Chinese *zhili xueshu fubai yu xueshu buduan de silu yu duice*]. *Social Science Forum* (Chinese journal *shehui kexue luntan*), (9), 30-63.

Juran, J. M., & Gryna, F. M., Jr. (Eds.). (1988). *Juran's quality control handbook* (4th ed.). New York: McGraw-Hill.

Keeling, R. P., & Hersh, R. H. (2011). *We're losing our minds: Rethinking American higher education*. New York: Palgrave Macmillan.

Kettunen, J. (2008). A conceptual framework to help evaluate the quality of institutional performance. *Quality Assurance in Education, 16*(4), 322-332.

Kezar, A. J. (2001). *Understanding and facilitating organizational change in the 21st century: Recent research and conceptualizations* (ASHE-ERIC Higher Education Report, Volume 28, Number 4). San Francisco: Jossey-Bass.

Kitagawa, F. (2003). New mechanisms of incentives and accountability for higher education institutions: Linking the regional, national and global dimensions. *Higher Education Management and Policy, 15*(2), 99-116.

Kitzinger, J. (1995). Qualitative research. Introducing focus groups. *British Medical Journal, 311*, 299-302.

Kolmos, A., Fink, K. F., & Krogh, L. (2004). The Aalborg model – problem-based and project-organized learning. In A. Kolmos, K. F. Fink & L. Krogh (Eds.), *The Aalborg PBL model – Progress, diversity and challenges* (pp. 9-18). Aalborg: Aalborg University Press.

Krippendorff, K. (2004). *Content analysis: An introduction to its methodology* (2nd ed.). Thousand Oaks, CA: Sage Publications.

Kvale, S., & Brinkmann, S. (2009). *InterViews: Learning the craft of qualitative research interviewing*. Thousand Oaks, CA: Sage.

Law, D. C. S. (2010). Quality assurance in post-secondary education: Some common approaches. *Quality Assurance in Education, 18*(1), 64-77.

Lee, J. C. K., Huang, Y. X., & Zhong, B. (2012). Friend or foe: The impact of undergraduate teaching evaluation in China. *Higher Education Review, 44*(2), 5-25.

Lee, W. O. (1996). The cultural context for Chinese learners: Conceptions of learning in the Confucian tradition. In D. A. Watkins & J. B. Biggs (Eds.), *The Chinese learner: Cultural, psychological, and contextual influences* (pp. 25-41). Hong Kong: Comparative Education Research Centre, Faculty of Education, University of Hong Kong; Melbourne: The Australian Council for Educational Research.

Leung, F. K. S. (1998). The implications of Confucianism for education today. *Journal of Thought, 33*, 25-36.

Levitt, T. (1972). Production-line approach to service. *Harvard Business Review, 50*(5), 41-52.

Li, F., John Morgan, W., & Ding, X. (2008). The expansion of higher education, employment and over-education in China. *International Journal of Educational Development, 28*(6), 687-697.

Li, Xiaoqun (2000). The expansion of higher education institutions and quality control [in Chinese *gaoxiao kuozhao yu zhiliang kongzhi*]. *Higher Education Exploration* (Chinese journal *gaojiao tansuo*), (2), 2-6.

Li, Y., Whalley, J., Zhang, S. & Zhao, X., 2008, *The Higher Educational Transformation of China and its Global Implications*. NBER Working Paper No. 13849. (Cambridge, Massachusetts, National Bureau of Economic Research).

Liefner, I. (2003). Funding, resource allocation, and performance in higher education systems. *Higher Education, 46*(4), 469-489.

Maaløe, E. (2009). *Coming to terms: Modes of interpretation, explanation and understanding in social research with particular reference to case studies*. Aarhus: Aarhus School of Business.

Meade, P., & Woodhouse, D. (2000). Evaluating the effectiveness of the New Zealand Academic Audit Unit: Review and outcomes. *Quality in Higher Education, 6*(1), 19-29.

Merriam-Webster Dictionary. (n.d.). quality. Retrieved May, 27 2010 from http://www.merriam-webster.com/dictionary/quality

Merton, R.K. (1957). *Social theory and social structure* (revised and enlarged edition). Glencoe, IL: Free Press.

Min, Weifang, (2002). Economic transition and higher education reform in China. Presentation note prepared for the *Higher Education Seminar at the Center on Chinese Higher Education* at Columbia University, January 24, 2002, New York, USA. Retrieved January 28, 2013, from
http://www.tc.columbia.edu/centers/coce/pdf_files/EconTransitionandHEReform.pdf

Min, Weifang. (2004). Historical perspectives and contemporary challenges: The case of Chinese universities. Retrieved March 5, 2010, from Center on Chinese Education, Teacher's College at Columbia University:
http://www.tc.columbia.edu/centers/coce/pdf_files/c8.pdf

Ministry of Education, China. (2000). *Statistics newsletter of educational development in 1999* [in Chinese *1999 nian quanguo jiaoyu shiye fazhan tongji gongbao*]. Retrieved April 11, 2012, from
http://www.moe.gov.cn/publicfiles/business/htmlfiles/moe/moe_633/200407/841.html

Ministry of Education, China (2004a). *Plan for undergraduate teaching evaluation in regular higher education institutions* [in Chinese *putong gaodeng xuexiao benke jiaoxue gongzuo shuiping pinggu fang'an*]. Beijing: Ministry of Education, China.

Ministry of Education, China (2004b). *Action plan of education innovation 2003-2007* [in Chinese *2003-2007 jiaoyu zhenxing xingdong jihua*]. Beijing: Ministry of Education, China.

Ministry of Education, China. (2008). *Introduction to 'Project 211'* [in Chinese *'211 gongcheng' jianjie*]. Retrieved December 17, 2012, from
http://www.moe.gov.cn/publicfiles/business/htmlfiles/moe/moe_846/200804/33122.html

Ministry of Education, China. (2009a). *Statistical newsletter of educational development in 2008* [in Chinese *2008 nian quanguo jiaoyu shiye fazhan tongji gongbao*]. Retrieved August 5, 2010, from
http://www.moe.gov.cn/publicfiles/business/htmlfiles/moe/moe_633/201005/88488.html

Ministry of Education, China. (2009b). *Undergraduate and short-cycle students distribution among different types of institutions* [in Chinese *putong ben, zhuanke fen xingzhi leibie xueshengshu*]. Retrieved August 5, 2010, from
http://www.moe.gov.cn/publicfiles/business/htmlfiles/moe/s4633/201010/109898.html

Ministry of Education, China. (2009c). *Regular and adult undergraduate and short-cycle students among different runners* [in Chinese *putong, chengren ben, zhuanke fen jubanzhe xueshengshu*]. Retrieved August 5, 2010, from
http://www.moe.gov.cn/publicfiles/business/htmlfiles/moe/s4633/201010/109897.html

Ministry of Education, China. (2009d). *Adult undergraduate and short-cycle students among different types of institutions* [in Chinese *chengren ben, zhuanke fen xingzhi leibie xueshengshu*]. Retrieved August 5, 2010, from
http://www.moe.gov.cn/publicfiles/business/htmlfiles/moe/s4633/201010/109901.html

Ministry of Education, China. (2009e). *Students at higher education institutions* [in Chinese *gaodeng xuexiao (jigou) xuesheng shu*]. Retrieved August 5, 2010, from
http://www.moe.gov.cn/publicfiles/business/htmlfiles/moe/s4633/201010/109880.html.

Ministry of Education, China. (2011). *Introduction to 'Project 985'* [in Chinese *'985 gongcheng' jianjie*]. Retrieved December 17, 2012, from
http://www.moe.gov.cn/publicfiles/business/htmlfiles/moe/s6183/201112/128828.html

Ministry of Education, China. (2012a). *Statistical newsletter of educational development in 2010* [in Chinese *2010 nian quanguo jiaoyu shiye fazhan tongji gongbao*]. Retrieved April 11, 2012 from
http://www.moe.gov.cn/publicfiles/business/htmlfiles/moe/moe_633/201203/132634.html

Ministry of Education, China. (2012b). Memorandum on comprehensively enhancing higher education quality [in Chinese *jiaoyubu guanyu quanmian tigao gaodeng jiaoyu zhiliang de yijian*]. Retrieved November 2, 2012, from
http://www.gov.cn/zwgk/2012-04/20/content_2118168.htm

Ministry of Education & Ministry of Finance, China. (2007). *Guideline on implementing 'Undergraduate Teaching Quality and Innovation Project'* [in Chinese *guanyu shishi 'gaodeng xuexiao benke jiaoxue zhiliang yu jiaoyu gaige gongcheng' de yijian*]. Retrieved December 17, 2012, from
http://www.moe.edu.cn/publicfiles/business/htmlfiles/moe/moe_1623/201001/xxgk_79761.html

Neave, G. (1988). On the cultivation of quality, efficiency and enterprise: An overview of recent trends in higher education in Western Europe, 1986-1988. *European Journal of Education, 23*(1/2), 7-23.

Neave, G. (2009). The evaluative state as policy in transition: A historical and anatomical study. In R. Cowen & A. M. Kazamias (Eds.), *International handbook of comparative education* (pp. 551-568). Dordrecht: Springer.

Newton, J. (2000). Feeding the beast or improving quality? Academics' perceptions of quality assurance and quality monitoring. *Quality in Higher Education, 6*(2), 153-163.

Newton, J. (2002). From policy to reality: Enhancing quality is a messy business (LTSN Generic Centre/The learning and teaching support network). Retrieved October 4, 2012 from
http://www.heacademy.ac.uk/assets/York/documents/resources/database/id158_From_policy_to_reality_enhancing_quality_is_a_messy_business.rtf

Nilsson, K. A., & Wahlen, S. (2000). Institutional response to the Swedish model of quality assurance. *Quality in Higher Education, 6*(1), 7-18.

OECD. (2006). Higher education: Quality, equity and efficiency. *IMHE Info*, July 2006. Retrieved May 3rd, 2012 from Program on Institutional Management in Higher Education at: www.oecd.org/edu/imhe/37126826.pdf

Oxford English Dictionary. (n.d.). quality. Retrieved May, 27 2010 from
http://dictionary.oed.com/cgi/entry/50194223?query_type=word&queryword=quality&fir

st=1&max_to_show=10&sort_type=alpha&result_place=1&search_id=kvAt-iy1gY0-11019&hilite=50194223

Parasuraman, A., Zeithaml, V. A., & Berry, L. L. (1985). A conceptual model of service quality and its implications for future research. *The Journal of Marketing, 8*(6), 41-50.

Perkin, H. (2007). History of universities. In J. J. F. Forest & P. G. Altbach (Eds.), *International handbook of higher education* (pp. 159-205). Dordrecht: Springer.

Perrow, C. (1970). *Organizational analysis: A sociological view*. London: Tavistock Publications.

Pfeffer, J., & Salancik, G. R. (2003). *The external control of organizations: A resource dependence perspective*. California: Stanford University Press.

Pollitt, C., & Bouckaert, G. (2004). *Public management reform: A comparative analysis*. Oxford: Oxford University Press.

Power, M. (1999). *The audit society: Rituals of verification*. Oxford: Oxford University Press.

Powney, J., & Watts, M. (1987). *Interviewing in educational research*. London: Routledge & Kegan Paul.

Quinn, L. & Boughey, C. (2009). A case study of an institutional audit: A social realist account. *Quality in Higher Education, 15*(3), 263-278.

Riccucci, N. M. (2001). The 'old' public management versus the 'new' public management: Where does public administration fit in? *Public Administration Review, 61*(2), 172-175.

Salter, B., & Tapper, T. (2002). The politics of governance in higher education: The case of quality assurance. *Political Studies, 48*(1), 66-87.

Schwarz, S., & Westerheijden, D. F. (Eds.). (2004). *Accreditation and evaluation in the European higher education area*. Dordrecht: Springer.

Scott, W. R. & Davis, F. G. (2007). *Organizations and Organizing: Rational, Natural, and Open System Perspectives*. New Jersey: Pearson Prentice Hall.

Shattock, M. (2010). *Managing successful universities*. Berkshire: Open University Press.

Sim, J. (2001). Collecting and analysing qualitative data: Issues raised by the focus group. *Journal of Advanced Nursing, 28*(2), 345-352.

Stensaker, B., Langfeldt, L., Harvey, L., Huisman, J., & Westerheijden, D. (2011). An in-depth study on the impact of external quality assurance. *Assessment & Evaluation in Higher Education, 36*(4), 465-478.

Sun, Yuanlei (2001). On education quality after the expansion of higher education institution [in Chinese *guanyu 'gaoxiao kuozhao' hou jiaoyu zhiliang de sikao*]. *Modern University Education* (Chinese journal *xiandai daxue jiaoyu*), (5), 96-98.

Tang, M., & Zuo, X. (2004). Advice to double the students enrolled in higher education institutions [in Chinese *kuoda gaoxiao zhaoshengliang yibei de jianyi*]. Retrieved January 28, 2013, from http://finance.sina.com.cn/review/20041023/15201102716.shtml

Thompson, J. D. (1967; 2003). *Organizations in Action: Social Science Bases of Administrative Theory*. New York: McGraw-Hill (1967); New Brunswick, NJ: Transaction Publishers (2003).

Ulrich, W. (2001). The quest for competence in systemic research and practice. *Systems Research and Behavioral Science, 18*(1), 3-28.

van Damme, D. (2001). Quality issues in the internationalization of higher education. *Higher Education, 41*(4), 415-441.

van der Wende, M. C., & Westerheijden, D. F. (2001). International aspects of quality assurance with a special focus on European higher education. *Quality in Higher Education, 7*(3), 233-245.

van Kemenade, E., & Hardjono, T. W. (2010). A critique of the use of self-evaluation in a compulsory accreditation system. *Quality in Higher Education, 16*(3), 257-268.

van Vught, F.A. and Westerheijden, D.F. (1993). *Quality management and quality assurance in European higher education: Methods and mechanisms.* Luxembourg: EU Commission.

van Vught, F.A., & Westerheijden, D.F. (1994). Towards a general model of quality assessment in higher education. *Higher Education, 28*(3), 355-371.

Vaughn, S., Schumm, J. S., & Sinagub, J. M. (1996). *Focus group interviews in education and psychology.* Thousand Oaks, CA: Sage.

Vieira, F. (2002). Pedagogic quality at university: What teachers and students think. *Quality in Higher Education, 8*(3), 255-272.

Volkwein, J. F., Lattuca, L. R., Harper, B. J., & Domingo, R. J. (2007). Measuring the impact of professional accreditation on student experiences and learning outcomes. *Research in Higher Education, 48*(2), 251-282.

Vroeijenstijn, A. I. (1999). The international dimension in quality assessment and quality assurance. *Assessment & Evaluation in Higher Education, 24*(2), 237-247.

Wang, Xufang. (2011). 2011 China social sciences blue book: 3 reasons for the lack of academic master [in Chinese *2011 nian <shehui kexue lanpishu>: chansheng buliao xueshu sixiang dashi de yuanyin you san*]. Retrieved February 5, 2013, from China Social Sciences Net at: http://www.cssn.cn/news/413911.htm

Weimer, D. L., & Vining, A. R. (2005). *Policy analysis: Concepts and practice.* Upper Saddle River, New Jersey: Prentice Hall.

Wellington, J. (2000). *Educational research: Contemporary issues and practical approaches.* London: Continuum.

Woodhouse, D. (2003). Quality improvement through quality audit. *Quality in Higher Education, 9*(2), 133-139.

Yang, Ming, & Chen, Wengan (2002). The impacts of higher education expansion on the quality of higher education [in Chinese *gaoxiao kuozhao dui gaodeng jiaoyu de yingxiang*]. *Researches in Higher Education of Engineering* (Chinese journal *gaodeng gongcheng jiaoyu yanjiu*), (3), 49-51.

Yin, R. (2009). *Case study research: Design and methods* (4th ed). Thousand Oaks, California: Sage.

Zhang, Chengbin. (2005). Challenges for intervening in the psychological crisis of college students [in Chinese *gaoxiao xuesheng xinli weiji ganyu mianlin de tiaozhan*]. *Education Exploration* (Chinese journal *jiaoyu tansuo*), (2), 102-103.

Zhang, Chunping (2006). On the quality of higher education in the period of mass higher education [In Chinese *lun dazhonghua jieduan de gaodeng jiaoyu fuwu zhiliang*]. *Heilongjiang Education (Higher Education Research & Appraisal),* (Chinese journal *Heilongjiang jiaoyu (gaojiao yanjiu yu pinggu)*), (7, 8), 18-20.

Zhong, B., Zhou, H., Liu, Z., & Wei, H. (2009). Study the background and trends, summarize the experiences, innovate thoughts on evaluation – a discussion of basic issues for a new round of undergraduate teaching evaluation [in Chinese *yanjiu beijing qushi, zongjie jingyan jiaoxun, chuangxin pinggu silu – xinyilun benke jiaoxue pinggu jiben wenti tanxi*]. *China Higher Education* (Chinese journal *zhongguo gaodeng jiaoyu*), (1), 31-34.

Zhou, Ji. (2007). Implementing 'Quality Project' and 'The No. 2 Policy', enhancing higher education quality all-roundly – speech at video conference on implementing 'Undergraduate Teaching Quality and Innovation Project' [in Chinese *shishi 'zhiliang* gongcheng' guancheng '2 hao wenjian' quanmian tigao gaodeng jiaoyu zhiliang]. Retrieved October 23, 2012, from the Quality Project website: http://www.zlgc.org/Detail.aspx?Id=1140

Zhou, Xueguang. (2003). *Ten lectures on the sociology of organizations* [in Chinese *zuzhi shehuixue shijiang*]. Beijing: Social Science Document Publishing House.

Zou, Y., Du, X., & Rasmussen, P. (2012). Quality of higher education: Organisational or educational? A content analysis of Chinese university self-evaluation reports. *Quality in Higher Education, 18*(2), 169-184.

About the author

Yihuan Zou is currently an assistant professor at School of Education, Central China Normal University. He holds a MA in education from Beijing Normal University and a PhD from Aalborg University, Denmark. He has been involved in a variety of research projects in higher education and educational policies. And his research interests covers higher education (teaching and learning, organizational behavior of higher education institutions, policy), learning theories, social research methods, sociology of education, education in different cultures, etc. He is also interested in humanities and social science in general.

Lightning Source UK Ltd.
Milton Keynes UK
UKOW02n0843151214

243137UK00001B/52/P